RESIDUE REVIEWS

VOLUME 76

RESIDUE REVIEWS

Residues of Pesticides and Other
Contaminants in the Total Environment

Editor
FRANCIS A. GUNTHER

Assistant Editor
JANE DAVIES GUNTHER

Riverside, California

VOLUME 76

SPRINGER-VERLAG
NEW YORK   HEIDELBERG   BERLIN
1980

Coordinating Board of Editors

Foreword

Worldwide concern in scientific, industrial, and governmental communities over traces of toxic chemicals in foodstuffs and in both abiotic and biotic environments has justified the present triumvirate of specialized publications in this field: comprehensive reviews, rapidly published progress reports, and archival documentations. These three publications are integrated and scheduled to provide in international communication the coherency essential for nonduplicative and current progress in a field as dynamic and complex as environmental contamination and toxicology. Until now there has been no journal or other publication series reserved exclusively for the diversified literature on "toxic" chemicals in our foods, our feeds, our geographical surroundings, our domestic animals, our wildlife, and ourselves. Around the world immense efforts and many talents have been mobilized to technical and other evaluations of natures, locales, magnitudes, fates, and toxicology of the persisting residues of these chemicals loosed upon the world. Among the sequelae of this broad new emphasis has been an inescapable need for an articulated set of authoritative publications where one could expect to find the latest important world literature produced by this emerging area of science together with documentation of pertinent ancillary legislation.

The research director and the legislative or administrative advisor do not have the time even to scan the large number of technical publications that might contain articles important to current responsibility; these individuals need the background provided by detailed reviews plus an assured awareness of newly developing information, all with minimum time for literature searching. Similarly, the scientist assigned or attracted to a new problem has the requirements of gleaning all literature pertinent to his task, publishing quickly new developments or important new experimental details to inform others of findings that might alter their own efforts, and eventually publishing all his supporting data and conclusions for archival purposes.

The end result of this concern over these chores and responsibilities and with uniform, encompassing, and timely publication outlets in the field of environmental contamination and toxicology is the Springer-Verlag (Heidelberg and New York) triumvirate:

Residue Reviews (vol. 1 in 1962) for basically detailed review articles concerned with any aspects of residues of pesticides and other chemical contaminants in the total environment, including toxicological considerations and consequences.

Bulletin of Environmental Contamination and Toxicology (vol. 1 in
1966) for rapid publication of short reports of significant advances
and discoveries in the fields of air, soil, water, and food contamination and pollution as well as methodology and other disciplines
concerned with the introduction, presence, and effects of toxicants
in the total environment.

Archives of Environmental Contamination and Toxicology (vol. 1 in
1973) for important complete articles emphasizing and describing
original experimental or theoretical research work pertaining to the
scientific aspects of chemical contaminants in the environment.

Manuscripts for *Residue Reviews* and the *Archives* are in identical
formats and are subject to review, by workers in the field, for adequacy
and value; manuscripts for the *Bulletin* are not reviewed and are published
by photo-offset to provide the latest results without delay. The individual
editors of these three publications comprise the Joint Coordinating Board
of Editors with referral within the Board of manuscripts submitted to one
publication but deemed by major emphasis or length more suitable for
one of the others.

Coordinating Board of Editors

Preface

That residues of pesticide and other contaminants in the total environment are of concern to everyone everywhere is attested by the reception accorded previous volumes of "Residue Reviews" and by the gratifying enthusiasm, sincerity, and efforts shown by all the individuals from whom manuscripts have been solicited. Despite much propaganda to the contrary, there can never be any serious question that pest-control chemicals and food-additive chemicals are essential to adequate food production, manufacture, marketing, and storage, yet without continuing surveillance and intelligent control some of those that persist in our foodstuffs could at times conceivably endanger the public health. Ensuring safety-in-use of these many chemicals is a dynamic challenge, for established ones are continually being displaced by newly developed ones more acceptable to food technologists, pharmacologists, toxicologists, and changing pest-control requirements in progressive food-producing economies.

These matters are of genuine concern to increasing numbers of governmental agencies and legislative bodies around the world, for some of these chemicals have resulted in a few mishaps from improper use. Adequate safety-in-use evaluations of any of these chemicals persisting into our foodstuffs are not simple matters, and they incorporate the considered judgments of many individuals highly trained in a variety of complex biological, chemical, food technological, medical, pharmacological, and toxicological disciplines.

It is hoped that "Residue Reviews" will continue to serve as an integrating factor both in focusing attention upon those many residue matters requiring further attention and in collating for variously trained readers present knowledge in specific important areas of residue and related endeavors involved with other chemical contaminants in the total environment. The contents of this and previous volumes of "Residue Reviews" illustrate these objectives. Since manuscripts are published in the order in which they are received in final form, it may seem that some important aspects of residue analytical chemistry, biochemistry, human and animal medicine, legislation, pharmacology, physiology, regulation, and toxicology are being neglected; to the contrary, these apparent omissions are recognized, and some pertinent manuscripts are in preparation. However, the field is so large and the interests in it are so varied that the editors and the Advisory Board earnestly solicit suggestions of topics and authors to help make this international book-series even more useful and informative.

"Residue Reviews" attempts to provide concise, critical reviews of timely advances, philosophy, and significant areas of accomplished or needed endeavor in the total field of residues of these and other foreign chemicals in any segment of the environment. These reviews are either general or specific, but properly they may lie in the domains of analytical chemistry and its methodology, biochemistry, human and animal medicine, legislation, pharmacology, physiology, regulation, and toxicology; certain affairs in the realm of food technology concerned specifically with pesticide and other food-additive problems are also appropriate subject matter. The justification for the preparation of any review for this book-series is that it deals with some aspect of the many real problems arising from the presence of any "foreign" chemicals in our surroundings. Thus, manuscripts may encompass those matters, in any country, which are involved in allowing pesticide and other plant-protecting chemicals to be used safely in producing, storing, and shipping crops. Added plant or animal pest-control chemicals or their metabolites that may persist into meat and other edible animal products (milk and milk products, eggs, etc.) are also residues and are within this scope. The so-called food additives (substances deliberately added to foods for flavor, odor, appearance, etc., as well as those inadvertently added during manufacture, packaging, distribution, storage, etc.) are also considered suitable review material. In addition, contaminant chemicals added in any manner to air, water, soil or plant or animal life are within this purview and these objectives.

Manuscripts are normally contributed by invitation but suggested topics are welcome. Preliminary communication with the editors is necessary before volunteered reviews are submitted in manuscript form.

Department of Entomology   F.A.G.
University of California   J.D.G.
Riverside, California
July 23, 1980

Table of Contents

RESIDUE REVIEWS

VOLUME 76

Environmental and metabolic transformations of primary aromatic amines and related compounds

By George E. Parris*

Contents

---

* Division of Chemical Technology HFF-424, Bureau of Foods, Food and Drug Administration, 200 "C" St., S. W., Washington, D. C. 20204.

© 1980 by Springer-Verlag New York Inc.
Residue Reviews, Volume 76

I. Introduction

Nitrogen forms a variety of functional groups in combinations with carbon, hydrogen, and oxygen (Table I). These functional groups have been particularly useful for adapting and activating aromatic compounds for use as chemical intermediates in synthetic processes. In addition, many final products contain nitrogen functionalities including pesticides, explosives, drugs, dyes, antioxidants, and antiozonants (MEYLAN et al. 1976, MASON and SWEENEY 1976).

Occupational exposure to primary aromatic amines and related compounds has historically had the largest demonstrable impact on human health. Workers exposed to potent carcinogens such as benzidine and naphthylamine (VEYS 1972, ZAVON 1973, VAN ROOSMALEN et al 1979) have developed tumors of the bladder. More recently, the impact on consumers of food colors (DANIEL 1962, BROWN et al. 1978), cosmetics (SHAFER and SHAFER 1976), and fabric dyes (FOUSSEREAU 1972) has received increasing attention because of the large number of people exposed.

However, reports of local industrial pollution (GAMES and HITES 1977), agricultural soil contamination (SPROTT and CORKE 1971), and general environmental contamination are perhaps even more important. For example, a broad survey of surface waters by the Environmental Protection Agency (EPA) showed nitrobenzenes and nitrotoluenes at several sites (EWING et al. 1977). A similar survey of sewage sludge (ERICKSON and PELLIZZARI 1977) found chloroanilines in 4 of 9 cities sampled. In addition, the Food and Drug Administration (FDA) has found that aromatic amines (DIACHENKO 1979) and nitroaromatics (YURAWECZ 1978) can enter and have entered the food chain.

This review draws from many different disciplines in an attempt to summarize most of the research relevant to the environmental chemistry of primary aromatic amines and the related nitro and azo compounds. Because of the interdisciplinary nature of the review some references to basic text books are given. The focus is mainly on abiotic and microbial transformations of primary aromatic amines in soil, sediment, and water; however, some metabolic transformations by higher organisms are mentioned. The chemistries of secondary and tertiary amines are not discussed. This limitation means that important compounds such as diphenylamine, N-substituted phenylenediamines, and nitrosamines are not covered. The approach taken here concentrates on reaction types rather than on the fate of individual compounds.

II. Oxidation of anilines in soil and microbial cultures

In general terms, substitution of an amino group ($-NH_2$) for hydrogen ($-H$) on an aromatic ring increases the electron density of the delocalized "pi" electron system. Loss of electron density at nitrogen due to delocalization reduces the basicity of aromatic amines relative to aliphatic

Table I. *Some important nitrogen functional groups.*

Amino	$-NH_2, -NHR, -NR_2$
Ammonium	$-NH_x R^+_{4-x}$ $x = 1-4$
Amine oxide	$-\overset{+}{\underset{R}{N}}\overset{O^-}{\underset{R}{\diagdown}}$
Hydroxylamine	$-NHOH$
Amide	$-NH\overset{O}{\overset{\|}{C}}R$
Hydroxamic acid	$-\overset{OH}{\underset{}{N}}-\overset{}{\underset{O}{C}}R$
Imine (Anil, Shiff's base)	$-N = CR_2$
Oxime	$HO-N=CR$
Nitroso	$-N = O$
Nitrosamine	$-N-N = O$
Nitro	$-NO_2$
Azo	$-N = N-$
Azoxy	$-N = \underset{O}{\overset{}{N}}-$
Diazoamino (triazene)	$-N = N-NH-$
Hydrazo (hydrazene)	$-N-N-$
Hydrazone	$-N-N = CR_2$
Urea	$-NH\overset{O}{\overset{\|}{C}}NR_2$
Nitrone	$-\overset{O}{\overset{\|}{N}} = CR_2$

amines. However, loss of an electron from the valence orbitals of an aromatic amine results in a delocalized cation-radical so that aromatic amines are usually easier to oxidize than aliphatic amines. In fact, aromatic compounds such as phenylenediamine with two amino substituents on a benzene ring are spontaneously oxidized by molecular oxygen and ozone. This reaction accounts for the antioxidant and antiozonant properties of phenylenediamine and *N*-alkylated phenylenediamines, which make them important additives to vulcanized rubber products (HOLLIFIELD

1975). Less reactive anilines can be oxidized through enzymatic catalysis as will be discussed below.

a) The arylamino radical

Many of the products obtained from one-electron oxidations of aromatic amines can be rationalized through an understanding of the arylamino radical, $ArNH\cdot$:

$$Ar-NH_2 \rightarrow ArNH_2^{+\cdot} + e^- \rightarrow ArNH\cdot + e^- + H^+$$

The unpaired electron of the arylamino radical is delocalized to the *ortho-* and *para-*positions of the aromatic ring (LAND 1965). Delocalization stabilizes the nitrogen radical, but makes sites in the ring subject to reaction:

The stability of the radical (e.g., how long it will remain free in a particular environment) will depend upon the number and kinds of substituents on the ring. Substituents also determine the contribution of each resonance form to the average electron configuration of the radical and the steric accessibility of each reactive site.

If arylamino radicals are generated at high concentration relative to other reactive radicals in a reaction medium, they will combine with one another (BORDELEAU *et al.* 1972, IWAN *et al.* 1976, ROBERTS and CASERIO 1965 a, ENGELHARDT *et al.* 1977):

Three couplings are possible ($N\cdot+N\cdot$, $N\cdot+C\cdot$, and $C\cdot+C\cdot$) but the latter is not likely to occur unless the $N\cdot+N\cdot$ and $N\cdot+C\cdot$ mechanisms are inhibited by steric factors (ROBERTS and CASERIO 1965 a).

In the systems of interest to environmental chemists, the hydroxyl radical is apt to be the most important potential reaction partner available to the arylamino radical. Combination of the radicals $N \cdot + O \cdot$ yields arylhydroxylamine (SJOBLAD and BOLLAG 1977). Attack of the hydroxyl radical on the ring leads to phenolic products. For example, FLETCHER and KAUFMAN (1979) have detected *ortho*-hydroxylated products from the metabolism of chloroanilines by *Fusarium oxysporum* Schlecht:

Halogen is usually lost from halogenated anilines during metabolism. This reaction can also be postulated to result from hydroxyl addition to the ring (DANIELS and SAUNDERS 1953). Halogen can also be lost during other oxidation processes (see Section II d):

b) The action of oxidase and peroxidase enzymes

Enzymes which accept "hydrogen atoms" (*i.e.*, $e^- + H^+$) from various substrates occur widely in the biosphere (CANTAROW and SCHEPARTZ 1967 a, SAUNDERS *et al.* 1964). Some of these enzymes must have molecular oxygen as the electron acceptor and, drawing on the terminology of microbiologists, are called "oxygen-obligative" oxidases. Other oxidases are "facultative" and use either molecular oxygen (O_2) or hydroperoxide (HOOH) as the electron acceptor. Peroxidase enzymes require hydroperoxide as the electron acceptor. The nature of the substrate is usually not too critical as long as a labile proton and easily removable electron are available. Phenols are common, naturally occurring substrates and the enzymes which have evolved to oxidize phenols probably are the same ones which oxidize man-made (xenobiotic) anilines:

For example, SJOBLAD and BOLLAG (1977) found an oxidase from the soil fungus *Rhizoctonia praticola* which oxidized a number of phenols (including chlorophenols) and 4-methoxyaniline. Mono- and di-chloroanilines were not oxidized in the presence of this enzyme, possibly because of the electron-withdrawing effects of the chlorine substituents. Oxidase and peroxidase enzymes frequently occur extracellularly in soils and culture media (SJOBLAD and BOLLAG 1977, BORDELEAU *et al.* 1972, IWAN *et al.* 1976). Thus, substrates need not be able to penetrate live cells in order to be oxidized, and oxidation may occur under sterile conditions.

BORDELEAU *et al.* (1972) studied oxidation of chloroanilines by H_2O_2/ horseradish peroxidase type II in a flow system. They were able to observe transient intermediates including an arylhydroxylamine (ArNHOH). This intermediate can arise from a direct mixed-function oxidase activity of the enzyme or from a 2-step process in which the arylamino radical combines nonenzymatically with hydroxyl radical (Fig. 1). This ambiguity cannot be readily resolved.

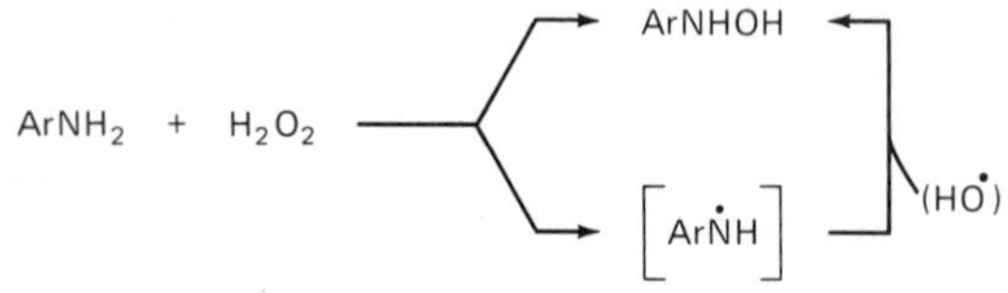

Fig. 1. Peroxidase oxidation of primary aromatic amine.

KAUFMAN *et al.* (1973) studied oxidation of 4-chloroaniline in cultures of the soil fungus *Fusarium oxysporum* Schlecht. They observed that this culture rapidly oxidized the arylamine to the arylhydroxylamine and the arylnitroso compound. Independently, it was shown that the arylhydroxylamine, arylnitroso, and arylnitro compounds rapidly interconverted to produce a mixture of products. These results demonstrate the reversibility of the oxidation and reduction reactions:

$$ArNH_2 \rightleftharpoons ArNHOH \rightleftharpoons ArNO \rightleftharpoons ArNO_2$$

In the aerobic medium, the equilibria favor the oxidized forms. Reduction of nitro functional groups will be discussed in Section III *b*.

BALBA *et al.* (1978) have recently studied incorporation of arylamine residues into synthetic lignins produced by copolymerization of coniferyl alcohol and arylamines with peroxidase systems (see Section IX *e*).

c) Quinone imines

While quinone imines have been isolated as metabolites of arylamines, it is important to note that they are not produced by the usual condensation route, $ArNH_2 + O = CR_2 \rightleftharpoons ArN = CR_2 + H_2O$. Quinone imines arise indirectly by the coupling and oxidation or disproportionation of arylamino radicals (BORDELEAU *et al.* 1972, VAN ALFEN and KOSUGE 1976, CORBETT 1969, MICHAELIS and HILL 1933):

The reaction between quinones and amines usually yields products with amino substituents on the quinoid ring (BIASAS 1974, HICKINBOTTOM 1957, CRANWELL and HAWORTH 1971):

d) The action of oxygenase and dioxygenase enzymes

Microorganisms attack aromatic rings with dioxygenase (WOOD et al. 1977) and monoxygenase (CERNIGLIA and GIBSON 1977) enzymes. The monoxygenase systems appear to function similarly to mammalian enzymes (KAUBISCH et al. 1972, OESCH 1972) (Fig. 2).

e) The action of chloroperoxidase enzymes

CORBETT et al. (1978 and 1979 a) have studied the oxidation of aromatic amines and hydroxylamines to the corresponding aromatic nitroso compounds by chloroperoxidase enzymes. Unlike peroxidase enzymes (see Section II b), chloroperoxidase enzymes catalyze 2-electron oxidations causing arylnitrenium ion-like rather than arylamino radical-like reactions.

While the chloroperoxidase enzyme isolated from the soil fungus *Caldariomyces fumago* oxidizes a number of ring-substituted anilines, aniline itself is the most rapidly oxidized. Substitution on the ring was found to inhibit formation of the aniline-enzyme complex, and hence slow down the rate of oxidation, for both electron-donating and electron-withdrawing substituents.

The effects of substituents appear to be mainly steric because no correlation between either K_m or V_{max} and redox potential, basicity or Hammett constants were found (CORBETT et al. 1979 a.).

It was suggested that, while oxidations of aromatic amines in soil are dominated by the abundance of peroxidase enzymes in that environment, oxidations of aromatic amines in the marine environment may be mainly caused by chloroperoxidases.

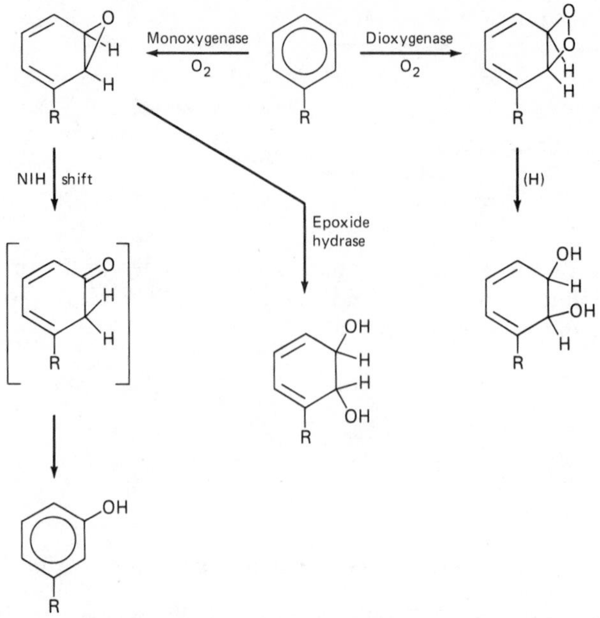

Fig. 2. Oxidation of aromatic rings (R = nitrogen functional group).

III. Reduction of nitro and azo groups

a) Reduction of nitrate

Inorganic nitrate ion is believed to be reduced to ammonia through a series of intermediates; $NO_3^- \rightarrow NO_2^- \rightarrow [HNO] \rightarrow H_2NOH \rightarrow NH_3$, in higher plants and microorganisms. The nitrate, nitrite, and hydroxylamine reductases are believed to be metalloflavoproteins requiring NADH or NADPH (BANKS 1968).

b) Reduction of nitro groups

Aromatic nitro groups are reduced enzymatically in several steps (YAMASHINA *et al.* 1954):

$$ArNO_2 \xrightarrow{H_2} ArNO + H_2O$$

$$ArNO \xrightarrow{H_2} ArNHOH$$

$$ArNHOH \xrightarrow{H_2} ArNH_2 + H_2O$$

Three moles of hydrogen are required for the process. McCORMICK *et al.* (1976) studied the relative rates of hydrogen uptake by aromatic nitro compounds in a reaction mixture containing cell-free extracts of *Veillonetta alkalescenes*. The rates depend upon the number, electronic effects, and substitution patterns of groups on the aromatic ring. Electron-withdrawing groups tend to make a nitro group more readily reducible. Aerobic as well as anaerobic systems were found to reduce nitro groups. The *V. alkalescens* "nitro-reductase" appears to involve a hydrogenase and ferredoxin-like material. It could not be determined whether or not ferredoxin acts as a nonspecific reductase for aromatic nitrocompounds. In the aerobic systems, "nitro-reductase" activity seems to involve NADH or NADPH and flavoproteins as in the reduction of inorganic nitrate.

It has been shown that the ability of mammals (*i.e.*, rats) to reduce ingested aromatic nitro compounds is mainly related to their gut microflora (WHEELER *et al.* 1975). Species of *Lactobacillus*, *Clostridium*, and *Streptoccocus* were found to increase the ability of initially germ-free rats to reduce *p*-nitrobenzoic acid to *p*-aminobenzoic acid.

c) Reduction of azo linkages

Numerous food coloring agents are azo dyes. Many workers have been concerned with the metabolism of these compounds with respect to cancer of the digestive system. It has been reported that the reduction of azo food dyes (as measured by color loss) was found to be zero kinetic order under *anaerobic* conditions in which the concentration of viable cells (e.g., *Proteus vulgaris*) remained constant (*i.e.*, the rates of reduction were independent of dye concentration). The zero-order rate constants correlated fairly well with redox potentials of the dyes. It was suggested that these facts can be explained by an extracellular, nonenzymatic process in which an electron carrier shuttles between the dye and intracellular reducing enzymes (Fig. 3) (DUBIN and WRIGHT 1975).

It was observed that cell mortality (with disruption of the cell membrane) caused an increase in reduction rate. Similar results were obtained with anaerobic cultures of *Bacteroides thetaiotaomicron* (CHUNG *et al.* 1978). In addition, it was found that the rates could be increased by addition of electron carriers, such as flavin mononucleotides or methyl viologen, to the culture media.

Analyses of effluents from dye manufacturing plants (GAMES and HITES 1977) suggest that reduction of azo dyes occurs during biological treatment of the manufacturing waste. The treatment system probably has anaerobic and aerobic microenvironments. Reduction of azo linkages in *aerobic* media has been studied by YONEZAWA and URUSHIGAWA (1977). These workers calculated the octanol/water partition coefficients for a group of azo dyes and related these results to growth inhibition of a culture of organisms from activated sludge. They observed that compounds which are more lipid-soluble (and hence can diffuse more readily

G. E. PARRIS

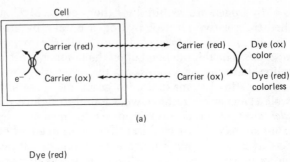

(a)

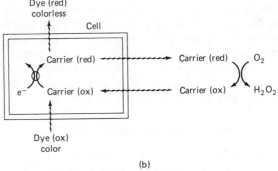

(b)

Fig. 3. Electron shuttle mechanism.

into cells) showed higher inhibition effects when compared at equal concentrations. They (URUSHIGAWA and YONEZAWA 1977) also determined the rate of discoloration of each dye. There was a positive correlation between rate of discoloration and growth inhibition. These results suggest that in an aerobic medium, the extra-cellular electron shuttle (Fig. 3) is quenched by oxygen and the rate of reduction is limited by the rate of diffusion of the azo compounds into the cells.

IV. Condensation of nitroso groups

The nitroso group is analogous to the aldehyde group and can undergo many similar reactions (SOLLENBERGER and MARTIN 1968). Since the nitroso group is an intermediate in the oxidation of amine groups and reduction of nitro groups as seen in the previous sections, it is not surprising that many condensation products are observed in metabolism studies. Azo and azoxy compounds are the most common products:

$$Ar-N=O + Ar-NH_2 \rightarrow Ar-N=N-Ar + H_2O$$

$$Ar-N=O + Ar-NHOH \rightarrow Ar-N=\overset{\overset{\displaystyle O}{|}}{N}-Ar + H_2O$$

The reaction involves two steps: In the first, a nucleophile attacks the nitroso nitrogen to yield an intermediate hydroxyhydrazine. Electron-withdrawing groups in the nitroso compound facilitate this step (OGATA and TAKAGI 1958):

$$\text{Ar—N=O} + \text{Ar—NHX} \rightarrow \overset{\displaystyle \text{OH}}{\underset{\displaystyle \text{X}}{\text{Ar—N—N—Ar}}}$$

The second step is fast and the products obtained depend on the nature of "X".

Since virtually every study of soil or microbial culture metabolism of aromatic amines (for example, KEARNEY and PLIMMER 1972, BARTHA and PRAMER 1967) or nitro compounds (for example, McCORMICK et al. 1978) reports condensation products, no attempt will be made to discuss each study. It is perhaps more interesting to note that *ortho* substituents inhibit formation of condensation products (BUSER and BOSSHARDT 1975).

Once formed, the condensation products seem to be relatively inert in soils. This fact, along with the bright colors of azo and azoxy compounds, probably accounts for their frequent isolation. Nonetheless, they should be susceptible to reduction (see Section III c). The 3,3',4,4'-tetrachloroazobenzene has gained attention because of its steric similarity to 2,3,7,8-tetrachlorodibenzodioxin (POLAND and GLOVER 1976, SPROTT and CORKE 1971).

V. Reactions involving environmental nitrite

Nitrite ion is produced in the environment by organisms which reduce nitrate and by other organisms which oxidize ammonia (FROBISHER et al. 1974). Thus, there is a constant turnover of inorganic nitrogen under aerobic and anaerobic conditions (SORENSEN 1978).

a) Diazonium formation

In a series of classic papers, HUGHES et al. (1958 a–1958 f) described the mechanism of diazonium formation:

$$H^+ + NO_2^- \rightleftharpoons HONO$$

$$H^+ + HONO \rightleftharpoons H_2ONO^+$$

$$H_2ONO^+ + NO_2^- \rightleftharpoons ONONO + H_2O \quad slow$$

$$ONONO + Ar—NH_2 \rightleftharpoons ArNH_2NO^+ + NO_2^-$$

$$ArNH_2NO^+ \rightleftharpoons ArN_2^+ + H_2O$$

Under most experimental conditions, formation of the anhydride of nitrous acid "ONONO" is the rate-limiting step in this process. The rate of that step (uncatalyzed) has been given as $k_2''[HNO_2]^2$ where $k_2'' = 0.8\ M^{-1}s^{-1}$ at ionic strength 0.1 M and 0 °C. This equation can be re-written:

$$\text{Uncatalyzed rate} = k_2'' \left[\frac{1}{K_a}[NO_2^-][H^+] \right]^2$$

where $K_a = 4.5 \times 10^{-4}\ M$ (HUGHES and RIDD 1958 c). Since pH (*i.e.*, $[H^+]$) can be measured directly, and at environmental levels of pH, $[NO_2^-]$free $>>$ $[HNO_2]$; $[NO_2^-]_{tot} \approx [NO_2^-]$free; it is convenient to to write

$$\text{Uncatalyzed rate} = \frac{0.8\ M^{-1}s^{-1}}{(4.5 \times 10^{-4}\ M)^2}\ [NO_2^-]^2_{tot}[H^+]^2$$

The rate of diazotization of any amine at any concentration cannot be faster than this rate unless the reaction is catalyzed by one of the mechanisms described by HUGHES *et al.* (1958 d).

The rates of reaction of various aromatic amines with nitrous anhydride have been determined and show a good correlation with the basicity of the amine (CHALLIS and BUTLER 1968). This reaction is similar to the condensation of an amine with a nitroso compound:

$$ArNH_2 + O{=}N{-}X \rightarrow Ar{-}N{=}N{-}X \rightleftharpoons Ar{-}N{\equiv}N^+ + X^-$$

The position of the final equilibrium depends upon the stability of the two ions and the solvent. There are many examples of various diazo compounds $Ar{-}N{=}N{-}X$ which are synthesized by coupling $Ar{-}N{\equiv}N^+$ with an appropriate ion X^- or HX:

$$ArN_2^+ + H_2NAr \rightarrow Ar{-}N{=}N{-}NH{-}Ar + H^+$$

$$ArN_2 + \text{⟨○⟩}{-}OH \rightarrow Ar{-}N{=}N{-}\text{⟨○⟩}{-}OH + H^+$$

b) Coupling reactions

Coupling of diazonium ions with phenols and other aromatic compounds has been a useful synthetic method (for example, CONANT *et al.* 1964) and the subject of many kinetic studies (CONANT and PETERSON 1930, WISTAR and BARTLETT 1941, HAUSER and BRESLOW 1941):

Successful coupling usually requires that the substituent "Z" be a strong electron-donating group.

The rates of coupling with phenols vary with pH. At low pH the re-action is inhibited because most of the phenol is in the protonated form. At high pH the phenolate ion is available, but the diazonium ion is converted to the diazotic acid ($ArN_2OH \rightleftharpoons ArN_2O^- + H^+$). Thus the rates are usually maximized near pH 7 where both the phenolate ion and the diazonium ion can exist (ROBERTS and CASERIO 1965 b).

CHISAKA and KEARNEY (1970) and PLIMMER et al. (1970) discovered and identified 1,3-bis(3,4-dichlorophenyl)triazene [i.e., N-(3,4-dichlorophenyldiazo)-3,4-dichloroaniline] among the metabolites of N-(3,4-dichlorophenyl)propionamide in soil. They determined that this product arose via free 3,4-dichloroaniline. These workers noted that the diazonium ion was a likely intermediate in the formation of the triazene. Work by MINARD et al. (1977) confirms that the role of microorganisms in this process is probably no more than incidental. They observed the transformation of 4-chloroaniline into the corresponding triazene in a culture of *Paracoccus sp.* The pH at which these processes occurred was 5.5 to 7.

c) Reduction of diazonium ions

CORKE et al. (1979) have observed formation of chlorinated biphenyls and azobenzenes from 3,4-dichloroaniline during microbial metabolism. They demonstrated that the diazonium ion was the precursor of these products by showing that only organisms which reduced nitrate to nitrite caused the conversion, and by stopping production of the products by trapping the intermediate with 2-naphthol.

These observations suggest a Meerwein-type reaction (RONDESTVEDT 1960) in which the diazonium is reduced by a one-electron reducing agent, such as chelated copper(I):

$$2ArN_2^+ + 2e^- \rightarrow Ar^{\cdot} + ArN_2^{\cdot} + N_2$$

It is known that under the Meerwein conditions the formation of symmetrical azo compounds is observed only when reaction between the azo compound and substrate (e.g., α, β-unsaturated ketone) is slow (FREUND 1951). MARCH (1968 a) notes that reduction to the biaryl is favored when ring substitutes are electron-withdrawing, while reduction to the azo-benzene is favored by electron-donating substituents.

It is not clear from CORKE's (1979) work whether or not enzymes are involved in the reduction or whether or not the reduction occurs inside the cells (see Section III c).

VI. Acylation of aromatic amines

Acylated metabolites of aromatic amines have frequently been observed. While acetyl derivatives are most often reported (TWEEDY et al. 1970, LU et al. 1977, BOLLAG and RUSSEL 1976, VISWANATHAN et al. 1978; BOLLAG et al. 1978), formyl (VISWANATHAN et al. 1978 a, KEARNEY and

PLIMMER 1972), and even propionyl (ENGELHARDT *et al.* 1977) metabolites have been observed. Acylation is regarded as a detoxification process for aromatic amines (CANTAROW and SCHEPARTZ 1967 b).

a) Acetyl metabolism

Acetyl and other acyl groups with two or more carbon atoms are transferred from one biochemical acceptor to another via coenzyme A (CoA) and appropriate transferase enzymes:

$$\text{Acetate} + \text{ATP} + \text{CoA} \rightarrow \text{Acetyl--CoA} + \text{AMP} + \text{PP}_i$$

$$\text{Acetyl--CoA} + \text{Acceptor} \rightarrow \text{Acetyl--Acceptor} + \text{CoA}$$

The acyl-CoA compounds are thioesters.

b) Formyl metabolism

The formyl group (—CHO) is one of several single carbon units (e.g., —CH$_3$, —CH$_2$OH) which are transported via tetrahydrofolic acid acting as a coenzyme.

It is noteworthy that one group (ANAGNOSTOPOULOS *et al.* 1978) has reported abiotic formation of N-(4-chlorophenyl)formamide in a sterile nutrient medium. Thus, the origin of formamides in soil metabolism studies should be checked with suitable controls.

c) Amidases

Aryl amides [e.g., N-(3,4-dichlorophenyl)propionamide] are readily hydrolyzed by microorganisms in soil but little information is available on this reaction (CHISAKA and KEARNEY 1970, BARTHA 1968 and 1971).

VII. Methylation of aromatic amines

N-Methylated metabolites of aromatic amines have been reported in a few environmental fate studies (e.g., LU *et al.* 1977). It is likely that S-adenosylmethionine and/or N^5-methyl-tetrahydrofolate are coenzymes in methylation processes.

Dealkylation of N-alkyl aromatic amines to produce primary amines has been observed (GOLAB *et al.* 1979). MACDONALD *et al* (1953) reported that demethylation of some N-methylamino aromatics occurs by oxidation to the N-hydroxymethyl compound, which is ultimately converted to formaldehyde and the amino compound. The aryl nitrone may be an intermediate in this oxidation (GORROD 1973).

VIII. Photochemistry of aromatic amines

BANERJEE *et al.* (1978) reported that photolysis of 3,3'-dichlorobenzidine (DCB) on water with natural sunlight causes rapid loss of chlorine and formation of monochlorobenzidine (MCB) and benzidine. The quantum yields for disappearance of DCB, MCB, and benzidine were determined at 3,000 and 2,537 Å. Little wavelength dependence was observed and the results were DCB 0.43, MCB 0.70, benzidine 0.012. These data are in agreement with the report (LU *et al.* 1977) that benzidine is rather slowly photolyzed in methanol.

Di-and mono-chlorobenzidines were found to be more stable in organic solvents (e.g., hexane, 2-propanol, methanol) than water, even though they are better hydrogen atom donors than water. It was also noted that low pH in aqueous solutions catalyzes the degradation of dichlorobenzidine. BANERJEE *et al.* (1978) concluded that the mechanism of dechlorination does not involve simple carbon-chlorine bond homolysis.

BANERJEE *et al.* (1978) observed that, at low pH, irradiated solutions of 3,3'-dichlorobenzidine and monochlorobenzidine (but not benzidine) develop a transient green color which decays in the dark. This phenomenon ocurs in oxygen-free solutions as well as in the presence of dissolved oxygen. The authors suggested a diphenoquinone-diimine type of intermediate as the transient and suggested that atomic or cationic chlorine is the oxidizing agent:

Indeed, they were able to generate similar color transients by adding chlorine water to solutions of dichlorobenzidine. The formation of the diphenoquinone-diamine may involve the cation radical of dichlorobenzidine obtained by one-electron oxidation:

accounting for the pH dependence of quantum yields discussed above (LAND 1965, LAND and PORTER 1963). LAND and PORTER (1963) pointed out that in water, 3 primary photochemical processes are possible:

$$ArNH_2 \rightarrow ArNH_2{}^+ + e^-$$

$$ArNH_2 \rightarrow Ar\dot{N}H + H\cdot$$

$$ArNH_3{}^+ \rightarrow Ar\dot{N}H_2{}^+ + H\cdot$$

while in paraffinic solvents only homolysis of the ArNH—H bond is possible.

The experiments with chlorine water (BANERJEE *et al.* 1978) are similar to work by JENKINS *et al.* (1978) on chlorination of aromatic amines in aqueous solvents. While anilines exposed to chlorine water tended to yield ring chlorination products, benzidine yielded a polymeric material without ring chlorination. The polymeric material is purple and the authors proposed a structure suggesting coupling of the type discussed in Section II *a* above.

IX. Covalent binding of anilines to soil organic matter

a) Sorption versus covalent binding

Covalent binding to soil organic matter should not be confused with sorption (BAILEY and WHITE 1970, KHAN 1972, PIERCE *et al.* 1971). Sorption involves formation of electrostatic, hydrophobic, and/or hydrogen bonds between a solute and some macromolecule with the collateral displacement of solvent (e.g., water) which may lead to a net positive entropy of sorption. These interactions are weak (typically $|\Delta H| \approx 10$ kcal/mole) (PIERCE *et al.* 1974, MOREALE and VAN BLADEL 1979) and the activation energies for exchange of free and sorbed molecules are found to be on the order of activation energies for viscous flow in the liquid phase (e.g., $|E_a| \approx 5$ kcal/mole) (BARROW 1966, BIGGAR *et al.* 1978).

The high rate of sorption/desorption reactions may explain the failure of WSZOLEK and ALEXANDER (1979) to observe desorption-rate-limited biodegradation of *n*-alkylamines sorbed to clay. The rate of biodegradation is probably limited by a second-order interaction between the amine and bacteria (PARIS *et al.* 1979):

$$\text{Amine·Clay} \overset{K_1}{\rightleftharpoons} \text{Amine} + \text{Clay} \quad (\text{fast})$$

$$\text{Amine} + \text{Bacteria} \overset{k_2}{\rightarrow} CO_2 \quad (\text{slow})$$

$$\text{rate} = \frac{d[CO_2]}{dt} = k_2[\text{Bacteria}][\text{Amine}]$$

$$= k_2[\text{Bacteria}]K_1 \frac{[\text{Amine·Clay}]}{[\text{Clay}]}$$

If the sorbant had been more porous (e.g., an organic soil rather than a clay) diffusion-rate-limited biodegradation might have been observed. For example, MOREALE and VAN BLADEL (1979) found that sorption of 4-chloroaniline by porous soil organic matter was described by an equation of the form

$$dq/dt = k(q_{max} - q_t)^n$$

where q is the concentration of sorbed amine at time t, q_{max} is the maximum concentration of sorbed amine, k is the rate constant, and n is the order of the reaction.

Unlike sorption, the position and dynamics of reversible covalent binding must be discussed in terms of specific mechanisms. The bonds formed and broken are inherently strong, but the free energy *change* for the overall process may be quite small. The mechanism of exchange may or may not involve high-energy transition states.

b) Composition of soil organic matter

Soil organic matter is usually classified according to its content of humic acids, fulvic acids, humin, etc. These fractions are operationally defined by their solubility in aqueous base and acid (SCHNITZER and KHAN 1972). Infrared spectra of these materials (STEVENSON and GOH 1971) and chemical analyses (RIFFALDI and SCHNITZER 1973) have revealed some of the functional groups which they contain. Common functional groups such as carboxyl, phenol OH, alcohol OH, carbonyl, and methoxyl usually are found at the milliequivalent/g level in humic acids.

c) Naturally-occurring nitrogen in humic acids

Humic acids normally contain 2 to 4% nitrogen (SCHNITZER and KHAN 1972). There are a number of theories as to the origin of this nitrogen (see references cited by BIEDERBECK and PAUL 1973). A significant portion of the nitrogen appears to be in the form of "humoproteins" (BIEDERBECK and PAUL 1973) which are acid hydrolyzable polypeptides bound (by an amino-terminal amino acid) to a humic moiety. These large complexes ($> 50{,}000$ M.W.) are hydrogen-bonded to other humic molecules containing less nitrogen.

The process by which polypeptides and lignin are converted to protein-lignin complexes during humification of plant and animal materials has been discussed by BRAUNS and BRAUNS (1960). CRANWELL and HAWORTH (1971) have studied model systems and suggest the reaction shown in Figure 4 as the mechanism by which terminal amino acids combine with humic quinone moieties. A primary amine can add to a quinone in a 1,2- or 1,4-fashion. Addition (1,2-) to the carbonyl yields a quinone imine, but this reaction is rapid and reversible. Water is a by-product and an excess prevents isolation of the quinone imine (CORBETT 1969). Addition (1,4-) to the α, β-unsaturated carbonyl may be slower than 1,2-addition and is also reversible. However, 1,4-addition is not inhibited by water and, more importantly, it leads to intermediates (by tautomerism) which can be irreversibly oxidized to an isolatable product. The isolation of quinone oximes and other 1,2-addition products under these reaction conditions appears to be due to favorable equilibria for their formation in certain situations.

d) Binding of anilines to humic acids

While a number of authors had noted that recovery of aromatic amines and their metabolites from soil metabolism studies was frequently far less than quantitative, HSU and BARTHA (1974 a and b) were the first

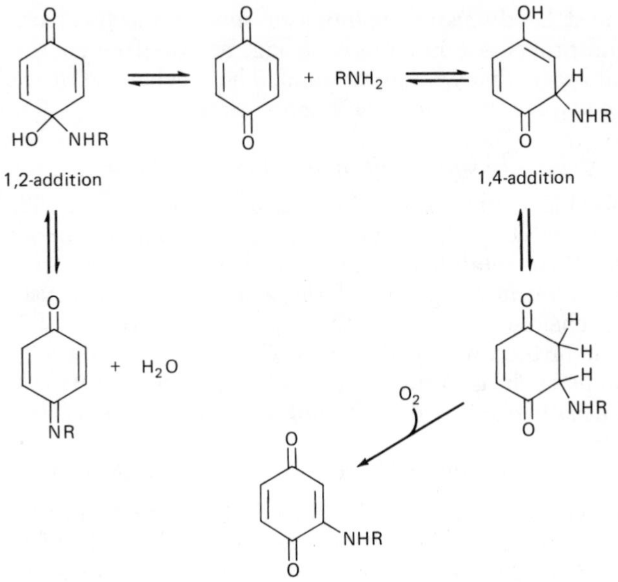

Fig. 4. Reactions of quinones with amines.

to seriously study the phenomenon. They recognized that some bound amines could be recovered by moderately vigorous hydrolysis, but that some amines were more or less irreversibly bound to the soil organic matter. They suggested imine (anil, Schiff's base) formation to account for reversible binding and several other reactions to account for non-hydrolyzable binding.

FUCHSBICHLER and SUSS (1978) found that radiolabeled 4-chloroaniline was extracted from soil more efficiently with solutions of unlabeled 4-chloroaniline or 3,4-dichloroaniline than with pure water, solutions of salt, or solutions of organic compounds not containing amino groups. This result suggests that amines are competitively bound to specific sites in the soil.

HELLING and KRIVONAK (1978 a; also see BALBA et al. 1979) studied the thermal stability of soil-bound residues of ^{14}C labeled dinitroanilines. Thermal analysis showed that most labeled CO_2 was evolved at temperatures below which the condensed core (CHESHIRE et al. 1967) or "nucleus" of soil organic matter decomposed. The thermal decomposition range suggested that the nitroaniline residues were ancillarily bound to carboxyl or phenol groups.

e) Incorporation of aromatic amines into lignin during plant growth

Historically, interest in the fate of the herbicide N-(3,4-dichlorophenyl)propionamide led to the conclusion that 3,4-dichloroaniline was the first metabolite produced by rice plants, but this compound was rapidly incorporated into the plant's lignin (STILL 1968, YAU et al. 1968). The analogy between phenol and aniline reactions with peroxidase/ H_2O_2 systems is discussed in Section II b of this review. BALBA et al. (1978) have adapted the peroxidase/H_2O_2 method for polymerization of coniferyl alcohol into synthetic lignin to make copolymers of coniferyl alcohol [3-(4-hydroxy-3-methoxyphenyl) allyl alcohol] and chloroanilines. They suggest that the copolymer has a C—N—C sigma bond at the C_1 position of the lignin polymer and that this results from 1,6-addition of the nitrogen to a quinone methide moiety. BALBA et al. (1979) have also described a method of analysis for these bound residues.

f) Biological activity of aniline residues bound to soil organic matter

Several authors have expressed concern about the biological activity of bound aniline residues. Humus-bound aromatic amine residues are not extractable in the normal methods used to monitor pesticides. Even if the humus-amine compounds were extractable, they might not be identifiable or quantifiable. Hence, there is potential for a large reservoir of undetectable and possibly toxic material to accumulate before the problem is recognized. Both the U.S. Environmental Protection Agency (EWING et al. 1977) and the U.S. Food and Drug Administration (i.e., LOMBARDO 1978) have made special efforts to avoid "surprise attacks" by previously unrecognized pollutants.

Little is known about the mechanisms of biodegradation of lignin or humic polymers. Some organisms appear to promote cleavage at the aliphatic sections of the polymer while other organisms attack the aromatic moieties before cleavage of the polymers. White-rot fungi are believed to attack the lignin polymers with a combination of extracellular mixed-function oxygenases and dioxygenases which mediate demethylations, hydroxylations, and ring-fission reactions (CRAWFORD and CRAWFORD 1979).

Hsu and BARTH (1974 b) found that humic-bound 3,4-dichloroaniline did not inhibit the overall rate of total CO_2 evolution by microbes. When the humic acid was used as the only organic substrate in a mineral salt medium for culture of *Penicillium frequentans* and *Aspergillus versicolor*, two modes of attack on the humic acid were demonstrated. Using ^{14}C-labeled bound residues of 3,4-dichloroaniline, it was observed that *P. frequentans* incorporated little radioactivity into its mycelium and released little [14]CO_2 compared to the total amount of CO_2 it evolved. On the other hand, *A. versicolor* incorporated much more radioactivity into its mycelium and evolved much more radiolabeled CO_2. It was concluded that *P. frequentans* attacks the humic acid, releasing oligomers with bound dichloroaniline moieties intact. *A. versicolor* mineralizes the aniline ring while it is still attached to humic acid (or immediately after it is removed from its attachment site). BOLLAG et al. (1978) recently reported additional work on binding and mineralization of chloroanilines.

Lichtenstein and coworkers (Lichtenstein *et al.* 1977, Katan and Lichtenstein 1977, Fuhremann and Lichtenstein 1978) published an interesting series of papers which emphasized the importance of distinguishing between mineralization (*i.e.*, conversion to CO_2, H_2O, etc.) and sequestration (e.g., covalent binding to macromolecules) for pesticides. They demonstrated that unextractable, soil-bound residues of methyl[ring-2,6-^{14}C] parathion (bound as aminoparathion) were taken up by earthworms and oat plants. On the other hand, Helling and Krivonak (1978 b) were frustrated in attempts to study the biological effects of bound dinitroaniline in soils using soybeans, because extraction of the soils to remove unbound residues created conditions which led to uptake of toxic levels of manganese. Viswanathan *et al.* (1978 b) reported data for soil binding and uptake of 3,4-dichloroaniline by barley in a long-term field study.

X. Metabolism in higher organisms

While metabolism of aromatic amines and related compounds by higher organisms may be outside the purview of the environmental chemist, some important processes are mentioned here.

a) Oxidation of nitrogen in organic compounds

Gorrod (1973) has reviewed oxidation of nitrogen in organic compounds by higher organisms (e.g., mammals). He concluded that 3 classes of compounds can be designated according to the pKa of the nitrogen: group I pKa > 7; group II pKa between 1 and 7; and group III pKa < 1. The basic amines of group I are oxidized by a flavine adenine nucleotide (FAD)-dependent enzyme system, while nonbasic group III compounds are oxidized by a cytochrome P450-dependent system. Compounds in group II are susceptible to both enzyme systems.

b) Conjugation of the amino group

N-Acylation (Kao *et al.* 1978) and N-methylation are common conjugation processes used by soil microorganisms to detoxify aromatic amines (see Sections VI and VII). There are many other conjugation reactions which higher plants and animals employ to prevent intoxication by endogenous and exogenous compounds (Smith and Williams 1966). However, many of these conjugates are quite labile and/or chemically complex so that they are difficult to identify in the environment or in tissues from organisms.

Glucuronic acid forms conjugates with a wide variety of compounds containing $-OH$, $-CO_2H$, $-SH$, and $-NH_2$ functional groups. Most of these conjugation reactions involve the uridine 5'-pyrophosphate-α-D-glucopyranosiduronic acid glucuronyl transferase (DUTTON 1966):

UDP = uridine 5'-pyrophosphate
Aglycon = $R-NH_2$, $R-OH$, $R-SH$, $R-CO_2H$
GT = glucaronyl transferase

However, amines may also form conjugates by a nonenzymatic pathway starting with the glucuronic acid (MARSH 1966):

glucuronic acid

H_2O ‖ $ArNH_2$

aryl-N-D-glucosiduronic
acid

The reverse of this reaction explains the acid-lability of amine conjugates of glucuronic acid. Arylamine-N-glucosiduronic acids have been detected in mammalian metabolism studies but they are not likely to be observed in the environment (BOYLAND et al. 1957). Similar mechanisms may be involved in formation of arylamine-N-glucosides with glucose as the conjugating agent.

Arylsulfamic acids are another common conjugate of aromatic amines (BRAY et al. 1956). They are formed by the transfer of SO_3 from 3'-phosphoadenosine-5'-phosphosulfate (PAPS) to the amino nitrogen with the aid of a sulfotransferase enzyme (DODGSON and ROSE 1970):

$$ArNH_2 + PA-OPO_2OSO_2O^- \rightarrow ArNHSO_3^- + PA-OPO_2O^- + H^+$$

Aromatic hydroxylamines (ArNHOH) form O-glucuronide, O-sulfate, and O-acetate conjugates by mechanisms similar to formation of the N-conjugates of aromatic amines (IRVING 1970).

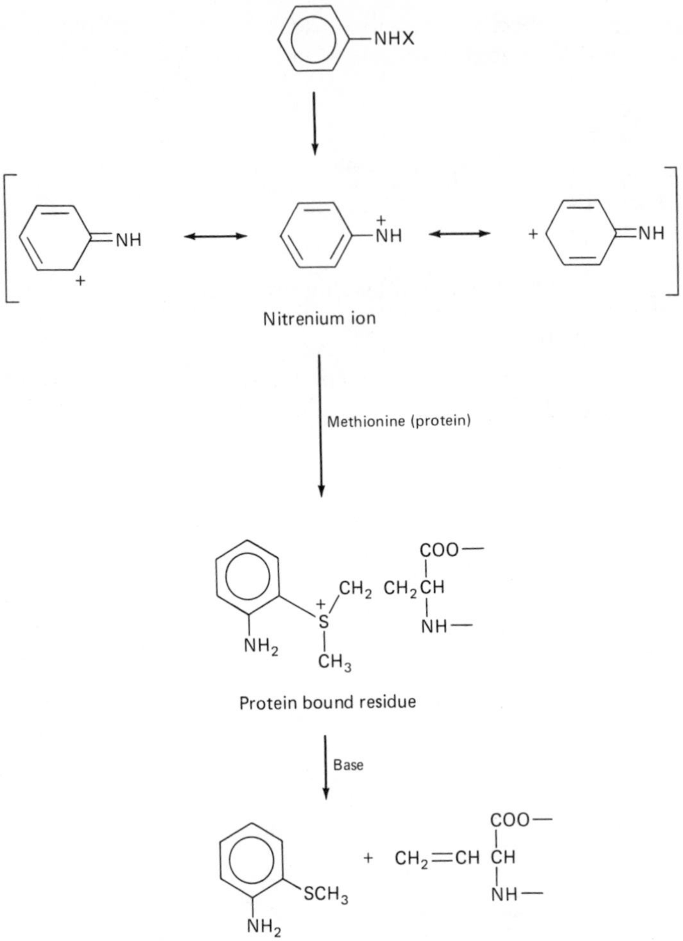

Fig. 5. Generation of thiomethyl derivatives of aromatic amines.

c) Covalent binding to macromolecules

Oxidation of aromatic amines *in vivo* produces arylnitrenium ion precursors and other compounds which have electrophilic properties. These electrophiles may conjugate with specially adapted cellular nucleophiles which inactivate them and/or facilitate their excretion. Some conjugations may be mediated by enzymes. The electrophiles also react (without enzymatic catalysis) with nucleophiles which are not normally electrophile scavengers (NEUMANN 1974). Three classes of macromolecules (*i.e.*, DNA, RNA, and protein) have been observed to combine with electrophilic aromatic amine residues. For example, HILL *et al.* (1979)

administered C^{14}-labeled 4-chloro-2-methylaniline to rats by intraperitoneal injection. After sacrificing the animals and compositing each organ type, they separated the DNA, RNA, and protein from each organ and analyzed the macromolecules for radioactivity. Most of the radioactivity was found in the liver where (in terms of radioactivity/unit weight) the order of bound residues was RNA > protein > DNA. The kidneys were also a major repository of bound residues, but little radioactivity was found in the muscle tissue.

Most arylnitrenium ion precursors have the formula ArNXR where *Ar* is an aryl moiety, *X* is a potential leaving group (OH, $OCOCH_3$, OSO_3H), and *R* can be any of several types of substituents (H, OR, $COCH_3$, alkyl, aryl). The nitroso compounds are a special case which can generate the nitrenium ion by protonation of the oxygen (KADLUBAR *et al.* 1978, MORTON *et al.* 1979, KRIEK 1969, MILLER *et al.* 1979, CORBETT *et al.* 1979 b). Substitution of a cellular nucleophile for the leaving group *X* can proceed by either an S_N1 or S_N2 type mechanism depending upon the solvent, the ability of *Ar* and *R* to delocalize positive charge, the stability of the leaving group *X*, the reactivity of the nucleophile, and the action of catalysts (MARCH 1968 b, KADLUBAR *et al.* 1978, IRVING 1970).

The specific nucleophiles which react with free, ion-paired, and incipient arylnitrenium ions have been discussed. Among the amino acids, NEUMANN (1974) and BARRY *et al.* (1969) pointed out the nucleophilic potential of methionine, cysteine, tyrosine, tryphtophan, and histidine. Methionine and cysteine are apparently the most active amino acid nucleophiles. Base hydrolysis of the amine-methionine adduct releases CH_3S-substituted aromatic amines (Fig. 5).

The purine and pyrimidine bases of RNA and DNA also react with electrophiles (NEUMANN 1974, KRIEK *et al.* 1967, KADLUBAR *et al.* 1978). KADLUBAR *et al.* (1978) have discussed the modes of binding of 1-naphthylamine residues to DNA and RNA. The major adduct is N-(deoxyguanosin-O^6-yl)-1-naphthylamine. A second adduct, 2-(deoxyguanosin-O^6-yl)-1-naphthylamine was also identified:

The authors speculated about how the presence of the naphthylamine moiety in DNA could induce (or accommodate) mis-pairing of guanine with thymine rather than with cytosine. Such mis-pairing is believed to lead to mutations and tumor formation (CLAYSON and GARNER 1976).

Summary

There are four points which should be made explicitly in a summary of the preceding discussions:

(1) It is not always clear from the published reports what the role of microorganisms is in the environmental transformations observed for aromatic amines and related compounds. Some reactions are catalyzed by enyzmes; but, in other cases, microorganisms only act to produce a local (intra- or extra-cellular) environment in which conditions are favorable for certain reactions.

(2) Classical organic chemists have explored many nonenzymatic reactions of nitrogen functional groups. This information should be a major source of inspiration and explanation for environmental scientists interested in the environmental fate of aromatic amines and related compounds.

(3) The high levels (e.g., 100 ppm) of aromatic amines used in most laboratory metabolism and fate studies prejudice the results seriously towards the formation of azobenzenes, bisaryl triazenes, and bisaryls from combination of two aromatic amine moieties. At more reasonable environmental levels (e.g., sub-ppm) and in the presence of potential reaction partners such as humic matter, it is likely that such products will be of minor importance.

(4) The interconversion of amino groups and the related azo-, azoxy-, nitro-, nitroso-, and amide groups appears to occur more rapidly than mineralization reactions. Thus, these compounds can be regarded as latent forms of aromatic amines in the environment.

References

ANAGNOSTOPOULOS, E., I. SCHEUNERT, W. KLEIN, and F. KORTE: Conversion of p-chloroaniline-^{14}C in green algae and water. Chemosphere 4, 351 (1978).

BAILEY, G. W., and J. L. WHITE: Factors influencing the adsorption, desorption, and movement of pesticides in soil. Residue Reviews 32, 29 (1970).

BAISAS, G. J.: The reaction of azidoquinones with nucleophiles and the chemistry of primary aminoquinones, Ph.D. thesis, order number 74-19, 241 (1974).

BALBA, H. M., G. G., STILL, and E. R. MANSAGER: Bound residue analysis of chloroaniline compounds in plants: A pyrolysis method. Presented Div. Pest. Chem. Amer. Chem. Soc., Anaheim, CA (Mar. 13, 1978).

—— —— —— Pyrolytic method for estimation of bound residues of chloroaniline compounds in plants. J. Assoc. Official Anal. Chemists 62, 237 (1979).

BANJERJEE, S., H. C. SIKKA, R. GRAY, and C. M. KELLY: Photodegradation of 3,3'-dichlorobenzidine. Environ. Sci. Technol. 12, 1425 (1978).

BANKS, B. E. C.: Biological formation and reactions of the amino group, pp. 517–519. In S. Patai (ed.): The chemistry of the amino group. New York: Wiley-Interscience (1968).

BARROW, G. M.: Physical Chemistry, 2nd Ed., pp. 544–545. New York: McGraw-Hill, (1966).

BARRY, E. J., D. MALEJKA-GIGANT, and H. R. GUTMANN: Interaction of aromatic amines with rat liver proteins in vivo. Chem-Biol. Interactions 1, 139 (1969).

BARTHA, R.: Biochemical transformations of anilide herbicides in soil. J. Agr. Food Chem. 16, 602 (1968).
—— Altered propanil biodegradation in temporarily air-dried soil. J. Agr. Food Chem. 19, 394 (1971).
——, and D. PRAMER: Pesticide transformation to aniline and azo compounds in soil. Science 156, 1617 (1967).
BIEDERBECK, V. O., and E. A. PAUL: Fractionation of soil humate with phenolic solvents and purification of the nitrigen-rich portion with polyvinylpyrrolidone. Soil Sci. 115, 357 (1973).
BIGGAR, J. W., U. MINGELGRIN, and M. W. CHEUNG: Equilibrium and kinetics of adsorption of picloram and parathion with soils. J. Agr. Food Chem. 26, 1306 (1978).
BOLLAG, J-M., P. BLATTMANN, and T. LAANIO: Adsorption and transformation of four substituted anilines in soil. J. Agr. Food Chem. 26, 1302 (1978).
——, and S. RUSSEL: Aerobic versus anaerobic metabolism of halogenated anilines by a Paracoccus sp. Microbial Ecol. 3, 65 (1976).
BORDELEAU, L. M., J. D. ROSEN, and R. BARTHA: Herbicide-derived chlorobenzene residues: Pathway of formation. J. Agr. Food Chem. 20, 573 (1972).
BOYLAND, E., D. MANSON, and S. F. D. ORR: The biochemistry of aromatic amines. Biochem. J. 65, 417 (1957).
BRAUNS, F. E., and D. A. BRAUNS: The chemistry of lignin, supplement volume 1949–1958, pp. 587–589. New York: Academic Press (1960).
BRAY, H. G., S. P. JAMES, and W. V. THORPE: Metabolism of the monochloronitrobenzenes in the rabbit. Biochem. J. 64, 38 (1956).
BROWN, J. P., G. W. ROEHM, and R. J. BROWN: Mutagenicity testing of certified food colors and related azo, xanthene and triphenylmethane dyes with the Samonella/microsome system. Mutation Res. 56, 249 (1978).
BUSER, H-R., and H.P. BOSSHARDT: Studies on the possible formation of polychloroazobenzenes in quintozene treated soil. Pest. Sci. 6, 35 (1975).
CANTAROW, A., and B. SCHEPARTZ: Biochemistry, pp. 347–365. Philadelphia: W. B. Saunders Co. (1967 a).
—— —— Biochemisty, pp. 280–281. Philadelphia: W. B. Saunders Co. (1967 b).
CERNIGLIA, C. E., and D. T. GIBSON: Metabolism of naphthalene by Cunninghamella elegans. Applied Environ. Microbiol. 34, 363 (1977).
CHALLIS, B. C., and A. R. BUTLER: Substitution at amino nitrogen, pp. 306–308. In S. Patai (ed.): The chemistry of the amino group. New York: Wiley-Interscience (1968).
CHESHIRE, M. V., P. A. CRANWELL, C. P. FALSHAW, A. J. FLOYD, and R. D. HAWORTH: Humic acid-II structure of humic acids. Tetrahedron 23, 1669 (1967).
CHISAKA, H., and P. C. KEARNEY: Metabolism of propanil in soils. J. Agr. Food Chem. 18, 854 (1970).
CHUNG, K-T., G. E. FULK, and M. EGAN: Reduction of azo dyes by intestinal anaerobes. Applied Environ. Microbiol. 35, 558 (1978).
CLAYSON, D. B., and R. C. GARNER: Carcinogenic aromatic amines and related compounds, pp. 366–461. In C. E. Searle (ed.): Chemical carcinogens. Washington, D. C.: American Chemical Society (1976).
CONANT, J. B., and W. D. PETERSON: The rate of coupling of diazonium salts with phenols in buffer solutions. J. Amer. Chem. Soc. 52, 1220 (1930).
——, R. E. LUTZ, and B. B. CORSON: 1,4-Aminonaphthol hydrochloride. Org. Syn. Collective Vol. I, 49 (1964).
CORBETT, J. F.: Benzoquinone imines. Part II. Hydrolysis of p-benzoquinone monoimine and p-benzoquinone diimine. J. Chem. Soc. B 1969, p. 213.
CORBETT, M. D., B. R. CHIPKO, and D. G. BADEN: Chloroperoxidase-catalysed oxidations of 4-chloroaniline to 4-chloronitrosobenzene. Biochem. J. 175, 353 (1978).
——, D. G. BADEN, and B. R. CHIPKO: Arylamine oxidations by chloroperoxidase. Bioorg. Chem. 8, 91 (1979 a).

—— —— —— The nonmicrosomal production of N-(4-chlorophenyl)glycohydroxa-
mic acid from 4-chloronitrosobenzene by rat liver homogenate. Bioorg. Chem. 8,
1 (1979 b).
CORKE, G. T., N. J. BUNCE, A. L. BEAUMONT, and R. L. MERRICK: Diazonium cations
as intermediates in the microbial transformation of chloroanilines to chlorinated
biphenyls, azo compounds, and triazenes. J. Agr. Food Chem. 27, 644 (1979).
CRANWELL, P. A., and R. D. HAWORTH: Humic acid-IV the reaction of alpha-amino
acid esters with quinones. Tetrahedron 27, 1831 (1971).
CRAWFORD, D. L., and R. L. CRAWFORD: Microbial degradation of lignin. Presented
Amer. Chem. Soc. Meeting, Sept. 12, Washington, D. C. (1979).
DANIEL, J. W.: The excretion and metabolism of edible food colours. Toxicol. Ap-
plied Pharmacol. 4, 572 (1962).
DANIELS, D. G. H., and B. C. SAUNDERS: Studies in peroxidase action. Part VIII. The
oxidation of p-chloroaniline. A reaction involving dechlorination. J. Chem. Soc.
1953, 822.
DIACHENKO, G. W.: Determination of several industrial aromatic amines in fish. En-
viron. Sci. Technol. 13, 329 (1979).
DODGSON, K. S., and F. A. ROSE: Sulfoconjugation and sulfohydrolysis, pp. 239–325.
In W. H. Fishman (ed.): Metabolic conjugation and metabolic hydrolysis, Vol. I.
New York: Academic Press (1970).
DUBIN, P., and K. L. WRIGHT: Reduction of azo food dyes in cultures of Protens
vulgaris. Xenobiotica 5, 563 (1975).
DUTTON, G. J.: The biosynthesis of glucuronides, pp. 185–300. In G. J. Dutton (ed.):
Glucuronic acid free and combined. New York: Academic Press (1966).
ENGELHARDT, G., P. WALLNÖFER, G. FUCHSBICHLER, and W. BAUMEISTER: Bacterial
transformations of 4-chloroaniline. Chemosphere 2/3, 85 (1977).
ERICKSON, M. D., and E. D. PELLIZZARI: Identification and analysis of polychlori-
nated biphenyls and other related chemicals in municipal sewage sludge samples.
Washington, D. C.: Environmental Protection Agency 560/6-77-021 (1977).
EWING, B. B., E. S. K. CHIAN, J. C. COOK, C. A. EVANS, P. K. HOPKE, and E. G.
PERKINS: Monitoring to detect previously unrecognized pollutants in surface
waters. Washington, D. C.: Environmental Protection Agency 560/7-77-001
(1977).
FLETCHER, C. L., and D. D. KAUFMAN: Hydroxylation of monochloroaniline pesticide
residues by Fusarium oxysporum schlecht. J. Agr. Food Chem. 27, 1127 (1979).
FOUSSEREAU, J.: Allergic eczema from disperse yellow 3 in nylon stockings and socks.
Trans. St. Johns Hospital Dermatol. Soc. 58, 75 (1972).
FREUND, W.: A new synthesis of arsonic acids, part I. Coupling of α, β-unsaturated
carbonyl compounds with diazotized p-arsonilic acid. J. Chem. Soc. 1951, 1943.
FROBISHER, M., R. D. HINSDILL, K. T. CRABTREE, and C. R. GOODHEART: Funda-
mentals of microbiology, pp. 669–673. Philadelphia: W. B. Saunders Co. (1974).
FUCHSBICHLER, G., and A. SÜSS: Desorption und austausch von sorbiertem 4-chlora-
nilin. Chemosphere 4, 345 (1978).
FUHREMANN, T. W., and E. P. LICHTENSTEIN: Release of soil-bound methyl [^{14}C]
parathion residues and their uptake by earthworms and oat plants. J. Agr. Food
Chem. 26, 605 (1978).
GAMES, L. M., and R. A. HITES: Composition, treatment efficiency, and environmental
significance of dye manufacturing plant effluents. Anal. Chem. 49, 1433 (1977).
GOLAB, T., W. A. ALTHANS, and H. L. WOOTEN: Fate of [^{14}C] trifluralin in soil. J.
Agr. Food Chem. 27, 163 (1979).
GORROD, D. W.: Differentiation of various types of biological oxidation of nitrogen in
organic compounds. Chem.-Biol. Interactions 7, 289 (1973).
HAUSER, C. R., and D. S. BRESLOW: Condensations. XV. The electronic mechanism
of the diazo coupling reaction. J. Amer. Chem. Soc. 63, 418 (1941).
HELLING, C. S., and A. E. KRIVONAK: Physiochemical characteristics of bound dinitro-
aniline herbicides in soil. J. Agr. Food Chem. 26, 1156 (1978 a).

—— —— Biological characteristics of bound dinitroaniline herbicides in soils. J. Agr. Food Chem. **26**, 1164 (1978 b).

HICKINBOTTOM, W. J.: Reactions of organic compounds, 3rd Ed., pp. 284–287. New York: Wiley (1957).

HILL, D. L., T.W. SHIH, and R. F. STRUCK: Macromolecular binding and metabolism of the carcinogen 4-chloro-2-methylaniline. Cancer Res. **39**, 2528 (1979).

HOLLIFIELD, H. C.: Alteration products of substituted p-phenylenediamines. Presented Conference on Environmental Aspects of Chemical Use in Rubber Processing Operations, Akron, Ohio, Feb. 7 (1975).

HSU, T-S., and R. BARTHA: Interaction of pesticide-derived chloroaniline residues with soil organic matter. Soil Sci. **116**, 444 (1974 a).

—— —— Biodegradation of chloroaniline-humus complexes in soil and in culture solution. Soil Sci. **118**, 213 (1974 b).

HUGHES, E. D., C. K. INGOLD, and J. H. RIDD: Nitrosation, diazotisation, and deamination. Part I. Principles, background, and method for the kinetic study of diazotisation. J. Chem. Soc. **1958 a**, 58.

—— —— —— Part II. Second and third order diazotisation of aniline in dilute perchloric acid. J. Chem. Soc. **1958 b**, 65.

——, and J. H. RIDD: Part III. Zeroth order diazotisation of aromatic amines in carboxylic acid buffers. J. Chem. Soc. **1958 c**, 70.

——, C. K. INGOLD, and J. H. RIDD: Part IV. Hydrogen ion catalysis in the diazotisation of o-chloroaniline in dilute perchloric acid. J. Chem. Soc. **1958 d**, 77.

——, and J. H. RIDD: Part V. Catalysis by anions of strong acids in the diazotisation of aniline and of o-chloroaniline in dilute perchloric acid. J. Chem. Soc. **1958 e**, 82.

——, C. K. INGOLD, and J. H. RIDD: Part VI. Comparative discussion of mechanisms of N- and O-nitrosation with special reference to diazotisation. J. Chem. Soc. **1958 f**, 88.

IRVING, C. C.: Conjugates of N-hydroxy compounds, pp. 53–119. In W. H. Fishman (ed.): Metabolic conjugation and metabolic hydrolysis, Vol. I. New York: Academic Press (1970).

IWAN, J., G-A. HOYER, D. ROSENBERG, and D. GOLLER: Transformations of 4-chloro-o-toluidine in soils: Generation of coupling products by one-electron oxidation. Pest. Sci. **7**, 621 (1976).

JENKINS, R. L., J. E. HASKINS, L. G. CARMONA, and R. B. BAIRD: Chlorination of benzidine and other aromatic amines in aqueous environments. Arch. Environ. Contam. Toxicol. **7**, 301 (1978).

KADLUBAR, F. F., J. A. MILLER, and E. C. MILLER: Guanyl O⁶-arylamination and O⁶-arylation of DNA by the carcinogen N-hydroxy-l-naphthylamine. Cancer Res. **38**, 3628 (1978).

KAO, J., J. FAULKNER, and J. W. BRIDGES: Metabolism of aniline in rats, pigs and sheep. Drug Metab. Distrib. **6**, 549 (1978).

KATAN, J., and E. P. LICHTENSTEIN: Mechanisms of production of soil-bound residues of [¹⁴C] parathion by microorganisms. J. Agr. Food Chem. **25**, 1404 (1977).

KAUBISCH, N., J. W. DALY, and D. M. JERINA: Arene oxides as intermediates in the oxidative metabolism of aromatic compounds. Isomerization of methyl-substituted arene oxides. Biochem. **11**, 3080 (1972).

KAUFMAN, D. D., J. R. PLIMMER, and U. I. KLINGBIEL: Microbial oxidation of 4-chloroaniline. J. Agr. Food Chem. **21**, 127 (1973).

KEARNEY, P., and J. R. PLIMMER: Metabolism of 3,4-dichloroaniline in soils. J. Agr. Food Chem. **20**, 584 (1972).

KHAN, S. U.: Adsorption of pesticide by humic substances. A review. Environ. Letters **3**, 1 (1972).

KRIEK, E.: On the mechanism of action of carcinogenic aromatic amines, I. Binding of 2-acetylaminofluorene and N-hydroxy-2-acetylaminofluorene to rat-liver nucleic acids *in vivo*. Chem.-Biol. Interactions **1**, 3 (1969).

——, J. A. MILLER, and E. C. MILLER: 8-(N-2-Fluorenylacetamido) guanosine, an arylamidation reaction product of guanosine and the carcinogen N-acetoxy-N-2-fluorenylacetamide in neutral solution. Biochem. 6, 177 (1967).

LAND, E. J.: Electronic spectra and kinetics of aromatic free radicals. In G. Porter (ed): Progress in reaction kinetics, vol. 3, pp. 394–399. New York: Pergamon (1965).

——, and G. PORTER: Primary photochemical processes in aromatic molecules, part 8. Absorption spectra and acidity constants of anilio radicals. Trans. Faraday Soc. 59, 2027 (1963).

LICHTENSTEIN, E. P., J. KATAN, and B. N. ANDEREGG: Binding of "persistent" and "nonpersistent" ^{14}C-labeled pesticides in an agricultural soil. J. Agr. Food Chem. 25, 43 (1977).

LOMBARDO, P.: FDA's chemical contaminants program: the search for the unrecognized pollutant. Presented Conference on Public Control of Environmental Health Hazards, New York, N. Y., June 27 (1978).

LU, P-Y., R. L. METCALF, N. PLUMMER, and D. MANDEL: The environmental fate of three carcinogens: Benzo-(a)-pyrene, benzidine, and vinyl chloride evaluated in laboratory model ecosystems. Arch. Environ. Contam. Toxicol. 6, 129 (1977).

MACDONALD, J. C., A. M. PLESCIA, E. C. MILLER, and J. A. MILLER: The metabolism of methylated aminoazo dyes. III. The demethylation of various N-methyl-C^{14}-aminoazo dyes in vivo. Cancer Res. 13, 292 (1953).

MARCH, J.: Advanced organic chemistry: Reactions, mechanisms, and structure, p. 555. New York: McGraw-Hill (1968 a).

—— Advanced organic chemistry: Reactions, mechanisms, and structure, pp. 281–302. New York: McGraw-Hill (1968 b).

MARSH, C. A.: Chemistry of D-glucuronic acid and its glycosides, pp. 3–136. In G. J. Dutton (ed.): Glucuronic acid free and combined. New York: Academic Press (1966).

MASON, R., and S. C. SWEENEY: A literature survey oriented towards adverse environmental effects resultant from the use of azo compounds, brominated hydrocarbons, EDTA, formaldehyde resins and o-nitrochlorobenzene. Washington, D. C.: Environmental Protection Agency 560/2-76-005 (1976).

McCORMICK, N. G., J. H. CORNELL, and A. M. KAPLAN: Identification of biotransformation products from 2,4-dinitrotoluene. Applied Environ. Microbiol. 35, 945 (1978).

——, F. E. FEEHERRY, and H. S. LEVINSON: Microbial transformation of 2,4,6-trinitrotoluene and other nitroaromatic compounds. Applied Environ. Microbiol. 31, 949 (1976).

MEYLAN, W. M., P. H. HOWARD, and M. SACK: Chemical market input/output analysis of selected chemical substances to assess sources of environmental contamination: Task I. Naphthylamines. Washington, D. C.: Environmental Protection Agency 560/6-77-002 (1976).

MICHAELIS, L., and E. S. HILL: Potentiometric studies on semiquinones. J. Amer. Chem. Soc. 55, 1481 (1933).

MILLER, E. C., F. F. KADLUBAR, J. A. MILLER, H. C. PITOT, and N. R. DRINKWATER: The N-hydroxy metabolites of N-methyl-4-aminoazobenzene and related dyes as proximate carcinogens in the rat and mouse. Cancer Res. 39, 3411 (1979).

MINARD, R. D., S. RUSSEL, and J-M. BOLLAG: Chemical transformations of 4-chloroaniline to a triazene in a bacterial culture medium. J. Agr. Food Chem. 25, 841 (1977).

MOREALE, A., and R. VAN BLADEL: Soil interactions of herbicide-derived aniline residues: A thermodynamic approach. Soil Sci. 127, 1 (1979).

MORTON, K. C., C. M. KING, and K. P. BAETCKE: Metabolism of benzidine to N-hydroxy-N,N'-diacetylbenzidine and subsequent nucleic acid binding and mutagenicity. Cancer Rec. 39, 3107 (1979).

NEUMANN, H-G.: Ultimate electrophilic carcinogens and cellular nucleophilic reactants. Arch. Toxikol. 32, 27 (1974).

OESCH, F.: Mammalian epoxide hydiases: Inducible enzymes catalysing the inactivation of carcinogenic and cytotoxic metabolites derived from aromatic and olefinic compounds. Xenobiotica 3, 305 (1972).

OGATA, Y., and Y. TAKAGI: Kinetics of the condensation of anilines wtih nitrosobenzenes to form azobenzenes. J. Amer. Chem. Soc. 80, 3591 (1958).

PARIS, D. F., W. C. STEEN, and G. L. BAUGHMAN: Kinetics of microbial transformations of pollutants in natural waters. Presented Amer. Chem. Soc. Meeting, Sept. 10, Washington, D. C. (1979).

PIERCE, R. H., JR., C. E. OLNEY, and G. T. FELBACK, JR.: Pesticide adsorption in soils and sediments. Environ. Letters 1, 157 (1971).

—— —— pp'-DDT adsorption to suspended particulate matter in sea water. Geochim. Cosmochim. Acta 38, 1061 (1974).

PLIMMER, J. R., P. C. KEARNEY, H. CHISAKA, J. B. YOUNT, and U. I. KLINGEBIEL: 1,3-Bis(3,4-dichlorophenyl)triazene from propanil in soils. J. Agr. Food Chem. 18, 859 (1970).

POLAND, A., and E. GLOVER: 3,4,3',4'-Tetrachloro-azoxybenzene and -azobenzene: Potent inducers of aryl hydrocarbon hydroxylase. Science 194, 627 (1976).

RIFFALDI, R., and M. SCHNITZER: Effect of 6N HCl hydrolysis on the analytical characteristics and chemical structure of humic acids. Soil Sci. 115, 349 (1973).

ROBERTS, J. D., and M. C. CASERIO: Basic principles of organic chemistry, pp. 915–916. New York: W. A. Benjamin (1965 a).

—— —— Basic principles of organic chemistry, pp. 892–895. New York: W. A. Benjamin (1965 b).

RONDESTVEDT, C. S., JR.: Arylation of unsaturated compounds by diazonium salts. Organic Reactions 11, 189 (1960).

SAUNDERS, B. C., A. G. HOLMES-SIEDLE, and B. P. STARK: Peroxidase. Washington, D. C.: Butterworths (1964).

SCHNITZER, M., and S. U. KHAN: Humic substances in the environment. New York: Marcel Dekker (1972).

SHAFER, N., and R. SHAFER: Potential of carcinogenic effects of hair dyes. N. Y. State J. Med. 76, 394 (1976).

SJOBLAD, R. D., and J-M. BOLLAG: Oxidative coupling of aromatic pesticide intermediates by a fungal phenol oxidase. Applied Environ. Microbiol. 33, 906 (1977).

SMITH, R. L., and R. T. WILLIAMS: Implications of the conjugation of drugs and other exogenous compounds, pp. 457–491. In G. J. Dutton (ed.): Glucuronic acid free and combined. New York: Academic Press (1966).

SOLLENBERGER, P. Y., and R. B. MARTIN: Carbon-nitrogen and nitrogen-nitrogen double bond condensation reactions, pp. 349–406. In S. Patai (ed.): Chemistry of the amino group. New York: Wiley-Interscience (1968).

SORENSEN, J.: Occurrence of nitric and nitrous oxides in a coastal marine sediment. Applied Environ. Microbiol. 36, 809 (1978).

SPROTT, G. D., and C. T. CORKE: Formation of 3,3',4,4'-tetrachloroazobenzene from 3,4-dichloroaniline in Ontario soils. Can. J. Microbiol. 17, 235 (1971).

STEVENSON, F. J., and K. M. GOH: Infrared spectra of humic acids and related substances. Geochim. Cosmochim. Acta 35, 471 (1971).

STILL, G. G.: Metabolism of 3,4-dichloropropionanilide in plants: the metabolic fate of the 3,4-dichloroaniline moiety. Science 159, 992 (1968).

TWEEDY, B. G., C. LOEPPKY, and J. A. Ross: Metobromuron: acetylation of the aniline moiety as a detoxification mechanism. Science 168, 482 (1970).

URUSHIGAWA, Y., and Y. YONEZAWA: Chemico-biological interactions in biological purification system II. biodegradation of azo compounds by activated sludge. Bull. Environ. Contam. Toxicol. 17, 214 (1977).

VAN ALFEN, N. K., and T. KOSUGE: Metabolism of the fungicide 2,6-dichloro-4-nitroaniline in soils, J. Agr. Food Chem. 24, 584 (1976).

VAN ROOSMALEN, P. B., A. L. KLEIN, and I. DRUMMOND: An improved method for determination of 4,4'-methylene bis-(2-chloroaniline) (MOCA)® in urine. Amer. Ind. Hyg. Assoc. J. 40, 66 (1979).

VEYS, C. A.: Aromatic amines: The present status of the problem. Ann. Occupational Hyg. **15**, 11 (1972).

VISWANATHAN, R., W. KLEIN, and F. KORTE: Separation and identification of metabolites excreted by rats after long-term oral administration of imugan-^{14}C. Chemosphere **1**, 71 (1978 a).

——, I. SCHEUNERT, J. KOHLI, W. KLEIN, and F. KORTE: Long-term studies on the fate of 3,4-dichloroaniline-^{14}C in a plant-soil-system under outdoor conditions. J. Environ. Sci. Health **B13**, 243 (1978 b).

WHEELER, L. A., F. B. SODERBERG, and P. GOLDMAN: The relationship between nitro group reduction and the intestinal microflora. J. Pharmacol. Expt. Therap. **195**, 135 (1975).

WISTAR, R., and P. D. BARTLETT: Kinetics and mechanisms of the coupling of diazonium salts with aromatic amines in buffer solutions. J. Amer. Chem. Soc. **63**, 413 (1941).

WOOD, J. M., R. L. CRAWFORD, E. MÜNCK, R. ZIMMERMAN, J. D. LIPSCOMB, R. S. STEPHENS, J. W. BROMLEY, L. QUE, JR., J. B. HOWARD, and W. H. ORME-JOHNSON: Structure and function of dioxygenases. One approach to lignin degradation. J. Agr. Food Chem. **25**, 698 (1977).

WSZOLOK, P. C., and M. ALEXANDER: Effect of desorption rate on the biodegradation of n-alkylamines bound to clay. J. Agr. Food Chem. **27**, 410 (1979).

YAMASHINA, I., S. SHIKATA, and F. EGAMI: Studies on enzymatic reduction of aromatic nitro, nitroso, and hydroxylamine compounds. Bull. Chem. Soc. Japan **27**, 42 (1954).

YAU, R. Y., D. H. McRAE, and H. F. WILSON: Metabolism of 3′,4′-dichloropropionanilide: 3,4-Dichloroaniline-lignin complex in rice plants. Science **161**, 376 (1968).

YONEZAWA, Y., and Y. URUSHIGAWA: Chemico-biological interactions in biological purification systems. I. Growth inhibition effect of azo compounds on activated sludge microorganisms. Bull. Environ. Contam. Toxicol. **17**, 208 (1977).

YURAWECZ, M. P.: GLC and mass spectrometric determination of monochloronitrobenzene residues in Mississippi River fish. Presented AOAC meeting, Oct. 17, Washington, D. C. (1978).

ZAVON, M. R.: Benzidine exposure as a cause of bladder tumors. Arch. Environ. Health **27**, 1 (1973).

Manuscript received December 21, 1979; accepted December 31, 1979

Conjugation of foreign chemicals by animals

By

G. D. PAULSON[*]

Contents

I. Introduction

Since the industrial revolution, and especially during this century, man and his environment have been exposed to an ever-increasing number of synthetic organic compounds. In many cases the compounds produced by our techology are, in a biological sense, "foreign chemicals" or "xenobiotics." That is, they are not synthesized by any known biological systems. The use of many of these compounds, which include a variety of drugs and pesticides, has had positive and far-reaching effects, both socially and economically. However, use of some of these foreign chemicals has had unexpected and, in some cases, undesirable effects. The

---

[*] Metabolism and Radiation Research Laboratory, Agricultural Research, Science and Education Administration, U. S. Department of Agriculture, Fargo, ND 58105. Mention of a trademark or proprietary product does not constitute a guarantee or warranty by the U. S. Department of Agriculture and does not imply its approval to the exclusion of other products that may be suitable.

awareness that the production and use of these compounds may have adverse effects on man and the environment has generated a great deal of interest concerning the fate of these compounds in biological systems. Because of this interest and because of the efforts of many investigators (particularly during the past 2 decades) there is now a large amount of published information concerning the metabolic fate of foreign chemicals in animals, including man (Fishman 1970; Lee *et al.* 1977; Williams 1947, 1959, and 1967; Hathway 1970, 1972, and 1975; Ladu *et al.* 1971; Parke 1968; Brodie and Gillette 1971; Parke and Smith 1977).

Williams (1947 and 1959) recognized and formalized the concept of phase I and phase II metabolism of foreign compounds. Phase I was defined as the initial oxidative, reductive, and hydrolytic metabolism of foreign compounds (Fig. 1). Products of phase I metabolism may be excreted without further metabolism. Often, however, phase I metabolites are further metabolized (conjugated) to yield phase II products (Fig. 1). When the foreign compound contains the appropriate functional group (e.g., $-OH$, $-COOH$, $-SH$, $-NH_2$), it may be conjugated directly. These type II reactions (conjugation reactions) which are catalyzed by a variety of transferase enzymes, are the subject of this review.

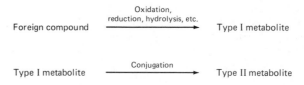

Fig. 1. The formation of Type I and Type II metabolites.

II. Historical aspects

The knowledge that animal systems can conjugate a variety of compounds is not new. Between 1842 and 1894, 8 of the 10 major conjugation reactions now known to occur in animals were identified (Williams 1947 and 1959, Smith and Williams 1970, Young 1977). The identification of most of the classes of conjugates was closely associated with the development of organic chemistry during the 19th century. Thus, it is not surprising that German organic and physiological chemists working during that era first recognized many of these reactions. Baumann (1876) is given credit for first using the term "gepaart" which means paired (or conjugated) in describing sulfuric acid conjugates as "gepaarte Schwefelsäure." Most of the conjugation reactions were first recognized by isolating the end products of metabolism from the urine of animals dosed with known compounds. However, the mechanisms involved in the biosynthesis of conjugates could not be elucidated until the techniques and procedures used to study intermediary metabolism were developed. Thus, our knowledge of the enzymes, cofactors, and substrates involved in the conjugation reactions is primarily the result of more recent studies by biochemists and physiologists. However, the work of metabolism chemists and pharmacologists has also contributed to our knowledge of biochemistry, particularly intermediary metabolism. For instance, study of the acetylation of sulfanilamide led to the discovery of acetyl CoA, and the investigation of glucuronic acid conjugation contributed to the discovery of the glucuronic acid pathway and the biosynthesis of vitamin C (SMITH and WILLIAMS 1970).

During the middle to late 19th century, a number of conjugated compounds were found to be much less toxic than the parent compound. The conjugates were also found to be more polar and therefore to be more efficiently eliminated in the urine and bile. These and other observations led to the consensus that conversion of toxic nonpolar compounds to polar conjugates was always a detoxication mechanism. Although this concept was generally accepted and promoted well into this century, more recent studies have revealed a number of exceptions. For instance, conjugation of certain N-hydroxy compounds, producing potent carcinogens, is an example of conjugation that produces a compound more deleterious than the parent molecule (IRVING 1970 and 1973, MILLER and MILLER 1969, WEISSBURGER et al. 1972, DEBAUN et al. 1970). Recent reports by RADOMSKI et al. (1977) indicated that the glucuronic acid conjugate of N-hydroxy-4-aminobiphenyl isolated from dog urine is mutagenic. DOROUGH (1976) reviewed other examples of compounds that were made more biologically active by conjugation. SMITH and WILLIAMS (1970) summarized the contemporary viewpoint on the biological significance of conjugation by stating "The general impression one gets from the study of conjugation products is that in the main they are relatively nontoxic, water soluble excretory products and their formation usually results in a reduction in toxicity although this process may not be perfect as might be expected in an imperfect world." As the methodology and instrumentation necessary for the isolation, identification, synthesis, and testing of conjugates improves, it seems inevitable that the biological significance of conjugation will be better defined.

III. General considerations

Logically it seems advantageous for an organism to be able to conjugate a wide variety of compounds and to have this conjugating ability in strategic locations. Animal systems generally meet these criteria in that they (1) can conjugate both nucleophilic and electrophilic centers and (2) have conjugating systems in the cytosol, the mitochondria, and the endoplasmic reticulum (Table I).

As is true of most anabolic processes, the conjugation of foreign compounds is energetically expensive. All of the conjugating systems except the one involving glutathione require the biosynthesis of a high-energy nucleotide, and the biosynthesis of glutathione also requires the expendi-

Table I. *Characteristics of major conjugation systems in animals.*

Type of conjugation	High-energy donor	Acceptor site conjugated	Site of conjugation reaction
Sulfate esterification	3'-Phosphoadenosine-5'-phosphosulphate (PAPS)	Nucleophilic center	Cytosol
Glucuronic acid	Uridine diphosphate glucuronic acid (UDPGA)	Nucleophilic center	Endoplasmic reticulum
Acetylation	Acetyl Coenzyme A (Ac CoA)	Nucleophilic center	Cytosol
Methylation	S-Adenosylmethionine (SAM)	Nucleophilic center	Cytosol and endoplasmic reticulum
Amino acid	Foreign compound—CoA	Activated foreign compound transferred to nucleophilic center of natural product (amino acid, etc.)	Mitochondria and endoplasmic reticulum and perhaps lysosomes
Glutathione and mercapturic acid	Intrinsic reactivity of foreign compound and glutathione (GSH); Ac CoA in last step of mercapturic acid formation	Electrophilic center	GSH conjugation in the cytosol; conversion to mercapturic acid in the endoplasmic reticulum

ture of high-energy compounds. Thus, all conjugation systems in animals require the expenditure of one or more high-energy equivalents (e.g., ATP, Acyl CoA). Therefore, these systems apparently have evolved because they are very important to the animal.

IV. Glucuronic acid conjugation

Glucuronic acid conjugation has been described as quantitatively the most important type of phase II metabolism (DUTTON et al. 1977, SMITH and WILLIAMS 1966) presumably because of the diverse functional groups that are conjugated with glucuronic acid and the relatively large amount of carbohydrate available for conjugation as compared to sulfate and amino acids, for example (DUTTON et al. 1977). The research leading to our present knowledge of the glucuronic acid pathway and the glucuronic acid conjugation of natural products and foreign compounds has been reviewed by WILLIAMS (1947 and 1959), ARTZ and OSMAN (1950), DUTTON (1966), MIETTINEN and LESKINEN (1970), and others.

The most common types of glucuronic acid conjugates fall into 4 general groups when classified on the basis of the functional group that is conjugated. The glucuronic acid moiety of the conjugates is linked: (1) through the oxygen of the hydroxyl group in phenols, alcohols (primary, secondary, and tertiary), and hydroxylamines (the O-ether type as shown in Figs. 2a, 2b, and 2c); (2) through the oxygen of the carboxyl group in carbocyclic and heterocyclic aromatic compounds, aliphatic acids, and aryl-alkyl acids (the O-ester types as shown in Figs. 2d and 2e); (3) through the nitrogen in anilines and some carboxyamides, sulfonamides, and heterocyclic compounds (the N-ether types as shown in Figs. 2f, 2g, 2h and 2i); or (4) through the sulfur of some thiols (S-ether types; Fig. 2j) and carbodithioic compounds (S-ester types; Fig. 2k). Extensive tabular listings of these 4 types of glucuronic acid conjugates have been prepared by DUTTON (1966), WILLIAMS (1947, 1959, and 1967), and others. The glucuronic acid moiety is bonded at the anomeric carbon and the linkage is in the β-configuration for all the glucuronic acid conjugates that have been identified.

More recent studies have demonstrated glucuronic acid conjugation of other functional groups. KADLUBAR et al. (1977) and RADOMSKI et al. (1977) reported a glucuronic acid conjugate of N-hydroxy-4-aminobiphenyl in which the linkage was with the nitrogen atom of the hydroxylamine group rather than with the oxygen atom (Fig. 2l). PORTER et al. (1975) and KENNEDY et al. (1978) reported a unique quaternary glucuronic acid conjugate of cyproheptadine, i.e., one with a carbon-nitrogen bond (Fig. 2m). SMITH et al. (1977) reported the direct glucuronidation of 5,5-diphenylhydantoin to a compound that was characterized as the N-3 glucuronide. LEVY et al. (1978) and YAGER et al. (1977) recently reported that Δ^6-tetrahydrocannabinol was converted to a C-glucuronide (Fig. 2n) by the mouse. This finding is especially interesting because the

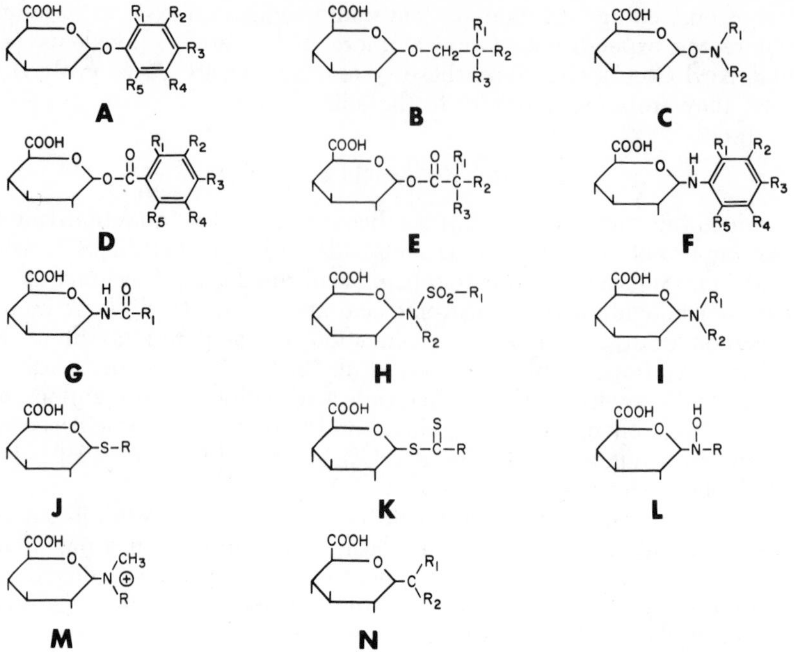

Fig. 2. Types of glucuronic acid conjugates formed in animal systems.

Δ^6-tetrahydrocannabinol molecule has a hydroxy group *ortho* to the site of C-glucuronic acid conjugation. RICHTER *et al.* (1975) and DIETERLE *et al.* (1976) previously reported the C-glucuronic acid conjugation of a group of pyrazolidine containing drugs (compounds containing a strongly acidic hydrogen on a carbon atom but no hydroxyl, amino, or thiol groups). Thus, the number of known types of glucuronic acid conjugates continues to grow.

The major pathway (perhaps the only quantitatively important one *in vivo*) for the biosynthesis of glucuronic acid conjugates is outlined in Figure 3. The conversion of uridine-5'-triphosphate (UTP) and glucose-1-phosphate to uridine-5'-diphoshate-α-D-glucose (UDPG) is catalyzed by a soluble enzyme (one that is in the cystosol) UDPG pyrophosphorylase (EC 2.7.7.9, UTP:α-D-glucose-1-phosphate uridyl transferase). Other quantitatively less important biosynthetic pathways leading to the formation of UDPG have been discussed by MIETTINEN and LESKINEN (1970). UDPG is oxidized to uridine-5'-diphosphate-α-D-glucuronic acid (UDPGA) under the influence of a microsomal enzyme, UDPG-dehydrogenase (EC 1.1.1.22, UDP glucose:NAD oxidoreductase). The conversion of UDP-iduronic acid to UDPGA has been described; however, it is probably not an important mechanism for the biosynthesis of UDPGA (MIETTINEN and LESKINEN 1970). UDPGA, the activated form of glucuronic acid, serves as the donor in conjugation reactions catalyzed by UDP-glucuronyl transferase (EC 2.4.1.17 UDP glucuronate glucuronyl transferase).

Because so many diverse compounds (both natural products and foreign compounds, with a great variety of functional groups) are conjugated with glucuronic acid, much attention has been devoted to the question of whether animal systems contain a single "glucuronyl transferase" or a number of glucuronyl transferase enzymes, each with a different degree of specificity. Much evidence supports the concept of multiple glucuronyl transferase enzymes. DUTTON and BURCHELL (1977) discussed a number of different types of heterogeneity possible for microsome-bound enzymes such as the glucuronyl transferase system; these include (1) sequential heterogeneity, (2) conformational heterogeneity (tertiary or quaternary protein structure), (3) occasional heterogeneity

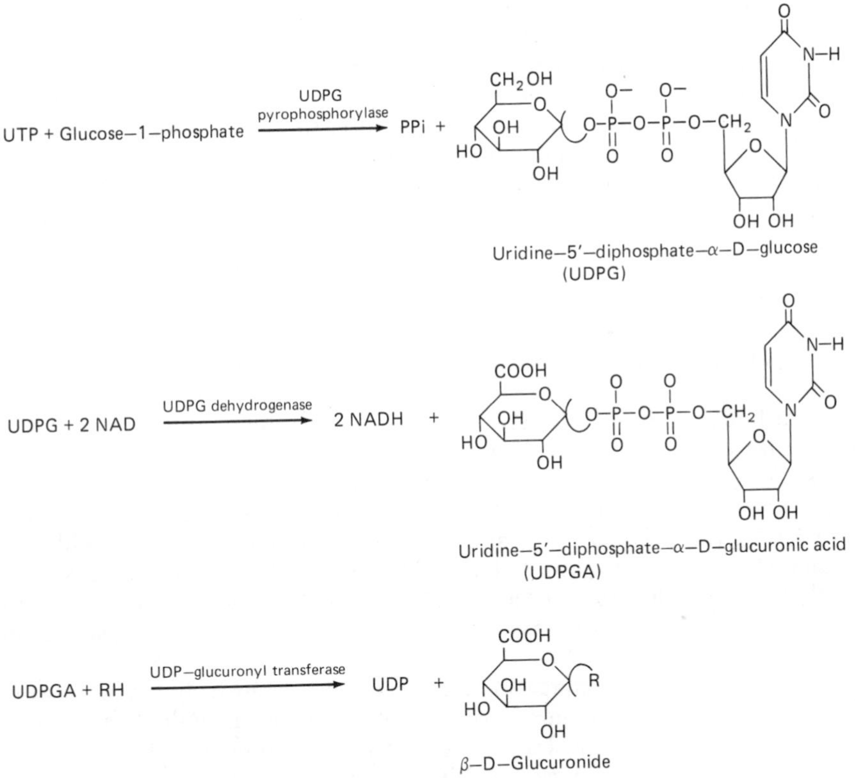

Fig. 3. The biosynthesis of uridine-5'-diphosphate-α-D-glucuronic acid (UDPGA) and glucuronic acid conjugates in animal systems.

(e.g. selective permeability of surrounding membranes to substrates), and (4) artifactual heterogeneity resulting from experimental techniques. Whether some or all of these mechanisms are involved in the apparent heterogeneity of the UDP glucuronyl transferase system(s) is still a matter of debate. Some workers have questioned the entire concept of heterogeneous forms of UDPG glucuronyl transferase enzymes; several reviews summarize the evidence for the opposing views (DUTTON 1966 and 1971, DUTTON and BURCHELL 1977, HUTSON 1970). The primary reason that this question has not been resolved is that a glucuronyl transferase enzyme has not been isolated. This failure is not surprising in light of the almost universally intimate association of the glucuronyl transferase activity with the endoplasmic reticulum (primarily in the rough microsomal fraction); numerous attempts to solubilize and purify the enzyme(s) have been plagued by the instability of the enzyme(s) and the difficulty of releasing the activity. Lacking a pure enzyme, researchers have had to rely on indirect methods of determining the homogeneity or heterogeneity of the glucuronyl transferase activity. This approach has generated a great deal of useful but, in some instances, conflicting and variously interpreted information. For instance, a number of "perturbation" procedures that "activate" UDP glucuronyl transferase have been described. Factors or conditions that may lead to activation include mechanical disruption, storage temperature, surfactants, solvents, sulfhydryl reagents, diethylnitrosamine, trypsin, phospholipases and phospholipids, fatty acids, UDP-N-acetylglucosamine and other UDP-sugars, reactants, and products. Some of these factors have been postulated as regulators for glucuronyl transferase activities *in vivo*. However, the question of which of these factors, if any, may be physiologically important is still subject to vigorous debate (DUTTON and BURCHELL 1977). Similarly, inhibition studies and other types of kinetics studies have yielded some results that are subject to various interpretations (DUTTON and BURCHELL 1977).

Although use of different techniques and different sources of the glucuronyl transferase enzyme(s) has sometimes yielded conflicting results, most indicate that the specificity of the enzyme(s) for UDPGA is quite high. DUTTON (1966) and DUTTON and BURCHELL (1977) have reviewed the information that indicates that other glucuronic acid-containing mucleotides (other than UDPGA), UDP-glucose, UDP-xylose, UDP-N-acetylglucosamine, and UDP galacturonic acid do not serve as substrates for the glucuronyl transferase enzyme(s) under physiological conditions.

Glucuronic acid conjugation has been most extensively studied in the liver; but it also occurs in a variety of other tissues, including the gastrointestinal tract, kidney, lung, spleen, thymus, adrenal gland, and skin (DUTTON et al. 1977, AITIO 1973, GRAM et al. 1974, HARTIALA 1973 and 1977). Although not all aspects of the process have been systematically studied in all species, there is ample evidence of glucuronide formation in mammals (WILLIAMS 1967, DUTTON *et al.* 1977, SMITH and CALDWELL 1977, DAVIES 1977), birds (WIT 1977), fish (SIEBER and ADAMSON 1977), and invertebrates (SMITH 1977). Exceptions such as the cat and the "gunn rat" have been reviewed by DUTTON (1971), SMITH (1968), and HIRAM *et al.* (1977).

DUTTON *et al.* (1977) and DUTTON and BURCHELL (1977) discussed reversability and transglucuronidation *in vitro*. For example, the glucuronic acid conjugation of *o*-aminophenol in the presence of UDP, *p*-nitrophenyl glucuronide, and liver microsomes has been observed. Knowledge about the importance of transglucuronidation, if any, *in vivo* must await further information on substrate concentrations *in vivo* (at the enzyme site) and factor(s) controlling the glucuronyl transferase enzyme(s).

It has been shown that β-glucuronidase (EC 3.2.1.31, β-D-glucuronide glucuronohydrolase) catalyzes the formation of glucuronic acid conjugates from free glucuronic acid and certain phenols *in vitro* (MIETTINEN and LESKINEN 1970). Other studies *in vitro* have demonstrated the β-glucuronidase-mediated transfer of glucuronic acid from aryl and alicyclic glucuronides to aliphatic alcohols and glycols but not to phenols or to alicyclic alcohols. However, MIETTINEN and LESKINEN (1970) summarized the evidence and concluded that "the direct role of β-glucuronidase in glucuronidation is negligible *in vivo*." A comprehensive review by WAKABAYASHI (1970) summarized and discussed the information available concerning the possible physiological role of β-glucuronidase.

The N-glucuronides of at least some classes of compounds can form nonenzymatically (DUTTON 1971). Whether this reaction is quantitatively important *in vivo* and how it may affect the overall metabolism of these classes of compounds are questions that apparently have not been pursued.

DUTTON and BURCHELL (1977), MILLBURN (1974), and HUNTER and CHASSEAUD (1976) reviewed many diverse factors that affect the formation of glucuronic acid conjugates. These include the animals species, sex, strain, age, and tissue locations of UDP-glucuronyl transferase; the route of administration of the foreign compound and its concentration at the site of conjugation; the availability of alternate conjugation mechanisms; and the animals' environment, diet, and exposure to other foreign compounds.

V. Other types of carbohydrate conjugation

It has been known for a number of years that some natural products may be conjugated with carbohydrate moieties other than glucuronic acid in animal systems. LAYNE (1970) reviewed the evidence for the conjugation of steroids with N-acetylglucosamine in rabbits and human beings and with glucose in rabbits. The linkage was at the anomeric carbon; usually the linkage was in the β-configuration, but in at least one case there was evidence for the linkage of the α-configuration.

LABOW and LAYNE (1974), GESSNER et al. (1973), and DUTTON et al. (1977) reported that mammalian microsomal preparations catalyze glucoside conjugation (in vitro with UDP-glucose as a substrate) of a variety of endogenous and foreign compounds, including steroids, retinol, bilirubin, p-nitrophenol, and diethylstilbestrol. Guinea pig microsomes, with UDP-galacturonic acid as the donor, catalyzed the galacturonic acid conjugation of p-nitrophenol in vitro (VESSEY and ZAKIM 1973). Kinetic studies indicated that the enzyme that catalyzed the galacturonic acid conjugation was distinct from the enzyme(s) responsible for glucuronic acid conjugation. Xylose conjugation of bilirubin in the presence of rat liver microsomes and UDP-xylose in vitro was reported by VAISMAN et al. (1976).

Thus, there is ample evidence that higher animals have the enzymatic ability to form glycoside conjugates other than glucuronic acid conjugates; and there is relevance to the questions of whether this ability is expressed in the intact higher animal and, if so, how quantitatively important are the other glycoside conjugations relative to glucuronic acid conjugation. It is apparent that, for many foreign compounds in many higher animal systems, glucuronic acid conjugation is the primary, if not the only, type of glycoside conjugation in vivo. However, there may be exceptions to this rule. For instance, DUGGAN et al. (1974) reported that the N-glucoside conjugate of 3-(4-pyrimidinyl)-5-(4-pyridyl)-1,2,4-triazole was a major metabolite of this compound in the bile and urine of dogs, rats, and monkeys. GESSNER and HAMANDA (1970) detected small amounts of p-nitrophenyl glucoside in the urine of mice given p-nitrophenol. The literature abounds with examples of "unidentified conjugates" and "presumed glucuronic acid conjugates." Further studies to determine the nature of the conjugating group(s) may lead to the identification of other glycoside conjugates.

VI. Sulfate ester and related conjugation

Although other classification systems have been suggested (DODGSON and ROSE 1970), there are 4 basic types of foreign compounds that are conjugated with a sulfate group by animal systems. These are phenols, alcohols, aromatic amines, and hydroxylamines (WILLIAMS 1967, IRVING 1970, DODGSON and ROSE 1970) (Fig. 4).

The early studies by BAUMANN (1876) (who first discovered the sulfate conjugation of phenol) and the studies by DeMEIO et al., BERNSTEIN and McGILVERY, BADDILEY, et al., and others who identified the cofactors and enzymes involved in sulfate ester biosynthesis have been reviewed by ROY (1960 and 1971), GREGORY and ROBBINS (1960), ROBBINS (1962), DeMEIO (1975), PAULSON (1976), and DODGSON and ROSE (1970). The sulfate is activated in a two-step reaction to give 3'-phosphoadenosine-5'-phosphosulfate (PAPS). The first step, catalyzed by the enzyme sulfate adenyl transferase (EC 2.7.7.4 ATP: sulfate adenyl transferase), involves

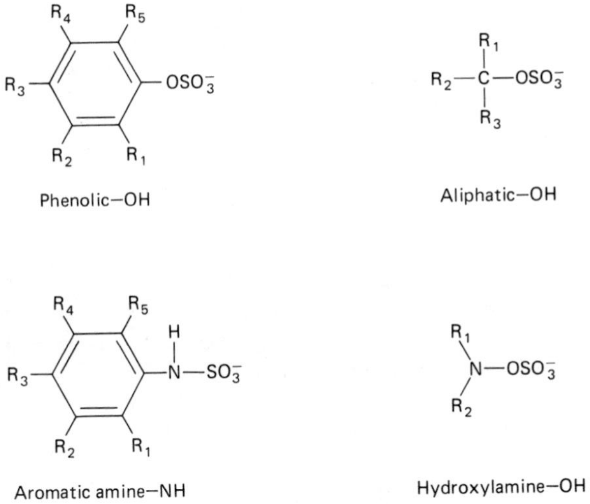

Fig. 4. Types of sulfate ester and related conjugates formed in animal systems.

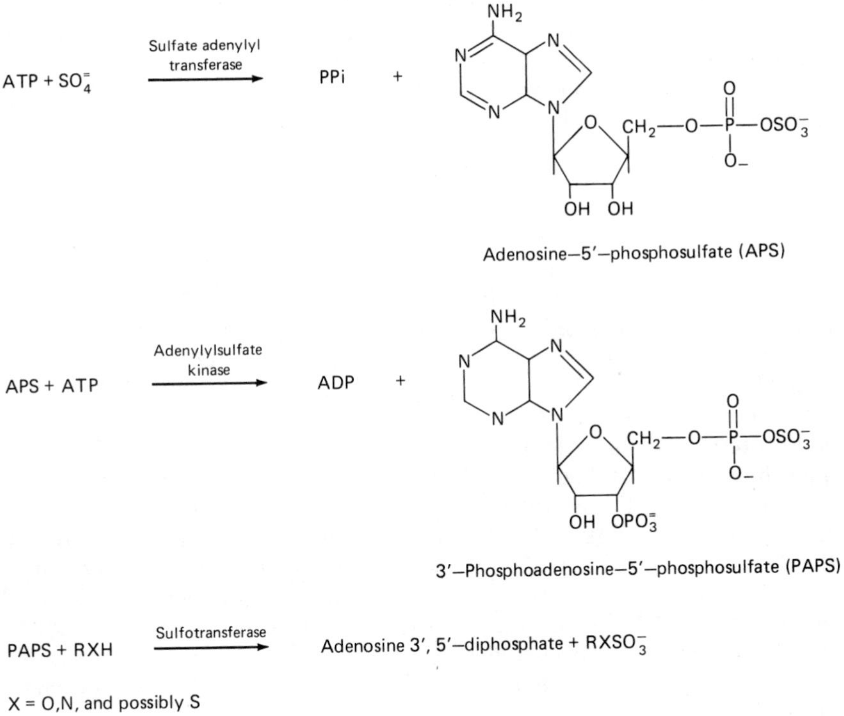

PAPS + RXH $\xrightarrow{\text{Sulfotransferase}}$ Adenosine 3', 5'–diphosphate + $RXSO_3^-$

X = O,N, and possibly S

Fig. 5. The biosynthesis of 3'-phosphoadenosine-5'-phosphosulfate (PAPS) and sulfate ester conjugates in animal systems.

the formation of adenosine-5'-phosphosulfate (APS) from ATP and SO_4^{2-} ion as shown in Figure 5. APS is phosphorylated under the influence of the enzyme adenyl sulfate kinase (EC 2.7.1.25, ATP:adenyl sulfate 3'-phosphotransferase) to give PAPS (activated sulfate) as shown in Figure 5. The enzymes involved in PAPS biosynthesis are in the cytosol. A comprehensive review on the occurrence, purification, and physical and kinetic properties of the sulfate adenyl transferase and adenyl sulfate kinase enzymes is available (PECK 1974).

The PAPS (activated sulfate) serves as the sulfate donor for the conjugation of many diverse natural products as well as foreign compounds. There is abundant evidence that animal systems contain a wide variety of sulfotransferase enzymes that catalyze the sulfate ester conjugation of natural products and foreign compounds.

The sulfate ester conjugation of phenols was the first sulfation process discovered, and subsequent studies have shown that a wide variety of foreign compounds is metabolized by this mechanism. Thus, aryl sulfotransferase (EC 2.8.2.1 3'-phosphoadenylylsulfate:phenol sulfotransferase) has received considerable attention. The aryl sulfotransferase activity is present in the particle-free supernatant (soluble fraction) of the cell types that have been systematically studied (WILLIAMS 1967, PAULSON 1976, DEMEIO 1975). Apparently no one has isolated an aryl sulfotransferase enzyme in pure form but the available information indicates that there is more than one form (DODGSON and ROSE 1970, PAULSON 1976, ROY 1960 and 1971, DEMEIO 1975).

The more recent information on the kinetic properties, substrate specificity, and other characteristics of some partially purified aryl sulfotransferase enzymes has been reviewed by DEMEIO (1975). There is direct and indirect evidence for aryl sulfate ester conjugation in a wide variety of animals including mammals, birds, amphibians, mollusks, insects, arachnids, and fish (SMITH 1955, 1968, and 1977; ROBBINS 1962; KOBAYASHI et al. 1975 and 1976; AKITAKE and KOBAYASHI 1975). Although sulfate ester formation has been most extensively studied in mammalian liver, there is little doubt it can occur in a variety of other tissues including the kidney, intestine, brain, adrenal, pancreas, mast cells, placenta, ovary, and testes (HARTIALA 1973 and 1977, ROY 1971, DEMEIO 1975). The quantitative importance of aryl sulfate conjugation varies with many factors including the animals' age (CARROL and SPENCER 1965, WENGLE 1963), species (DEMEIO 1945), and sex (MILLER et al. 1974); the size of the dose (MINCK et al. 1973, PAULSON and ZEHR 1971); sulfur nutritional state (SLOTKIN and DISTEFANO 1970, KURYZNSKE and SMITH 1975), and disease state of the animal (GESSNER 1974); time after dosing (KURYZNSKE and SMITH 1975); the occurrence of substituent effects (SATA et al. 1956); and the presence of inhibitors (MILLER et al. 1974, MULDER and PILON 1975).

In vitro studies have indicated that polyphenols can be polysulfated (VESTERMARK and BOSTRÖM 1960). However, polysulfation is apparently of little or no importance in the metabolism of foreign compounds in vivo

(DeMeio 1975). There is indirect evidence that thiophenols may be sulfoconjugated to form thiosulfates (Williams 1947 and 1959). However, this observation has apparently not been confirmed by the isolation and characterization of the proposed metabolites.

The conjugation of arylamines with sulfate to give the sulfamates is catalyzed by the soluble enzyme arylamine sulfotransferase (EC 2.8.2.3, 3'-phosphoadenylylsulfate:arylamine sulfotransferase). Although this enzyme has received relatively little attention, there is good evidence for its occurrence in the rat, rabbit, guinea pig, spider, and toad, as reflected by the conversion of simple arylamines such as aniline and 2-naphthylamines to the corresponding sulfamates in vivo (Williams 1967, Smith 1968, Roy 1971, DeMeio 1975). Small amounts of sulfamates are reported to be eliminated in the urine of rats and rabbits (Roy 1960, Boyland et al. 1957). However, Hollingsworth (1977) suggested that sulfamates are so unstable that they cannot be readily recovered from in vivo systems. Roy (1971) also discussed the inherent instability of some of these compounds. There may be notable exceptions, however; Williams (1947 and 1959) reviewed the evidence that sulfamic acid is stable when given to animals and is excreted intact in the urine and feces.

There is evidence for the formation of sulfate ester conjugates of aliphatic alcohols and polyhydroxy compounds in vitro and in vivo (Roy 1960 and 1971, Gregory and Robbins 1960, Gregory 1962). It seems certain that this conjugation is catalyzed by one or more sulfotransferase enzyme(s) and that it can occur in a wide variety of animals (Dodgson and Rose 1970, DeMeio 1975). However, research on the enzyme systems involved in this process has apparently been limited.

Earlier reports that the sulfate ester conjugates of certain hydroxylamine-containing compounds are excreted in the urine of animals have been challenged by Irving (1970) and others. Nevertheless, there is compelling evidence that sulfate ester conjugates of hydroxylamines are formed (both in vitro and in vivo), and the toxicological implications of some of these conjugates have generated much interest. There is little doubt that sulfate ester conjugation is involved in the conversion of certain "proximate carcinogens" to "ultimate carcinogens" (Irving 1970 and 1971, Miller and Miller 1969, Weissburger et al. 1972, Debaun et al. 1970). The most extensively studied compound of this class is 2-acetylamino fluorene (AAF). It has been shown that AAF is first oxidized to N-hydroxy-AAF by an oxygen dependent microsome enzyme(s) and the N-hydroxy AAF is then conjugated with glucuronic acid or with sulfate (Fig. 6). The glucuronide conjugate of N-hydroxy-AAF can be isolated from the urine of animals treated with AAF. To the author's knowledge, no one has isolated the sulfate ester of N-hydroxy-AAF from an in vivo system which is not surprising when one considers the highly reactive nature of this compound [Debaun et al. (1967) estimated that the half-life of this compound in water is less than 1 min]. Nevertheless, a compelling amount of information obtained by a series of ingenious in vitro and in vivo experiments supports the conclusion that the sulfate ester of

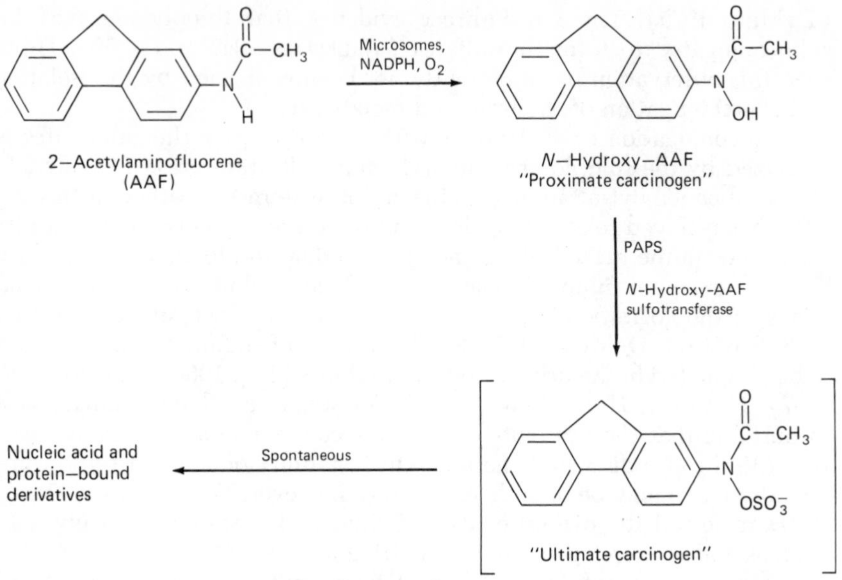

Fig. 6. The formation of the proximate and ultimate carcinogens from 2-acetylamino-
fluorene (AAF) in animal systems.

N-hydroxy-AAF (the "ultimate carcinogen") is formed *in vivo* (Irving
1970 and 1971, Miller and Miller 1969, Weissburger *et al.* 1972, Debaun
et al. 1967) (Fig. 6). The sulfate ester of N-hydroxy-AAF reacts spontaneously with protein, RNA, and DNA, leading to the covalent attachment of 2-acetylaminofluorene and 2-aminofluorene residues to these
macromolecules (Fig. 6) (Irving 1970 and 1971, Miller and Miller
1969, Weissburger *et al.* 1972, Debaun *et al.* 1970). Methionine, cysteine,
tryptophan, tyrosine, and guanine are the primary sites of attachment for
these residues in proteins and nucleic acids (Irving 1971). N-hydroxy-2-
acetyl aminofluorene transferase, the enzyme(s) responsible for the sulfate ester conjugation of N-hydroxy-AAF, has not been isolated. However,
it is a soluble enzyme (105,000 $\times$ g supernatant of rat liver homogenates),
and it requires PAPS (Irving 1970 and 1971, Miller and Miller 1969,
Debaun *et al.* 1967, King and Olive 1975).

A wide variety of aryl amine compounds is susceptible to N-hydroxylation (Irving 1970, Kiese 1966). There is evidence that at least some of
these N-hydroxy compounds are activated by sulfate conjugation. However, the structure of the aryl group has a strong effect on the nature of
the reactivity of the conjugated compound (Irving 1971, Miller and
Miller 1977). An interesting study by Blunck and Crowther (1975)
showed that the carcinogenic potential of 3'-methyl-4-dimethylaminoazo-
benzene in rats was markedly altered by factors affecting the liver sulfate

pool. Thus, sulfate ester conjugation also appears to have an important role in the carcinogenicity of this compound.

VII. Glutathione and mercapturic acid conjugation

The term "mercapturic acid" conjugates was coined by early investigators who observed that the polar urinary metabolites of some foreign compounds were degraded to thiophenols by basic hydrolysis. About 100 years ago the classic studies of BAUMANN and others demonstrated that S-(p-bromophenyl) acetylcysteine was present in the urine of animals dosed with bromobenzene. During that era the same type of mercapturic acid conjugation was demonstrated for other halogenated aromatic and related compounds. However, it was not demonstrated until the early 1960s that the cysteine moiety in these mercapturic acid conjugates is derived from glutathione (GSH). It is now well established that GSH and mercapturic acid conjugates are logically considered together because the latter are the result of further metabolism of GSH conjugates as shown in Figure 7. The discovery of mercapturic acid conjugates and the later studies leading to the present knowledge of GSH conjugation and the conversion of these compounds to mercapturic acid conjugates have been reviewed by BOYLAND and CHASSEAUD (1969), HUTSON (1970, 1972, 1975, and 1976), HOLLINGSWORTH (1977), GROVER (1977), BOYLAND (1962 and 1971), CHASSEAUD (1973 and 1976), and WOOD (1970). Exhaustive reviews of the biosyntheses and properties of GSH (the precursor to these two classes of compounds) are also available (MEISTER 1974 and 1975).

The conversion of a foreign compound to a mercapturic acid conjugate proceeds through a well-defined 4-step sequence (Fig. 7). The first reaction is catalyzed by a wide variety of GSH S-transferase enzymes and will be discussed later. Significant amounts of some compounds may react with GSH without enzyme catalysis, but there is ample evidence that these reactions are greatly enhanced by the GSH S-transferase enzymes. The second reaction, catalyzed by the enzyme γ-glutamyl transferase (EC 2.3.2.2), involves the removal of the glutamyl moiety from the GSH conjugate. The third step is the hydrolytic removal of the glycine moiety by the enzyme cysteinyl-glycine dipeptidase (EC 3.4.13.6) to yield the cysteine conjugate. The last step is the N-acetylation of the cysteine moiety by an acetyl CoA-requiring enzyme to yield the mercapturic acid conjugate. Although all the different GSH S-transferases are soluble enzymes (HOLLINGSWORTH 1977, BOYLAND 1971, CHASSEAUD 1973, WOOD 1970), it is interesting to note that all the enzymes responsible for the conversion of GSH conjugates to mercapturic acid conjugates (γ-glutamyl transferase, cysteinyl-glycine dipeptidase, and acetyl transferase) are membrane-bound enzymes (microsome fraction) (GROVER 1977, TATE et al. 1976).

Mercapturic acid conjugation has been observed in a wide variety of animals including the rabbit, rat, mouse, dog, cat, hamster, guinea pig,

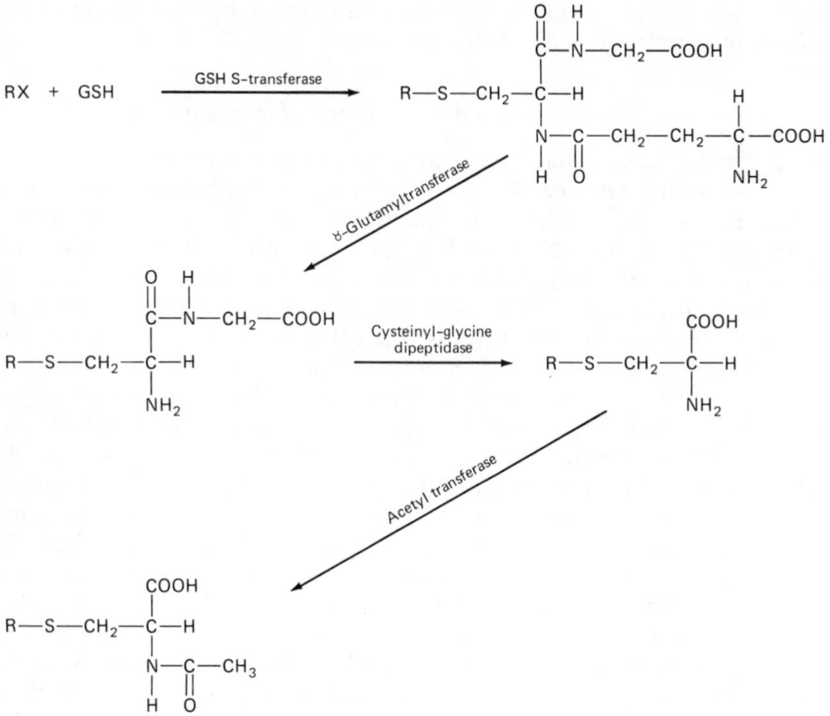

Fig. 7. The biosynthesis of glutathione (GSH) and mercapturic acid conjugates in animal systems.

horse, sheep, pig, monkey, cow, lizard, frog, chicken, duck, goldfish, and man (Wood 1970, Jakoby 1978). There is evidence for GSH S-transferase activity in diverse tissues including the lung, liver, kidney, and gut (Grover 1977). Furthermore, Jakoby (1978) has reviewed recent information that indicates that the intestine of the rat has very high GSH S-transferase activity. Although not all the intermediate products have always been studied systematically, there is evidence that the enzymatic ability to convert GSH conjugates to mercapturic acids varies with the tissue and species of animal (Boyland 1971).

As Chasseaud (1976) and others have discussed, GSH conjugates have the ideal physiochemical properties (negative charge and molecular weight exceeding 300) for preferential secretion in the bile. Thus, it is not surprising that the GSH conjugates, as well as the catabolites of the GSH conjugates (cysteinyl-glycine, cysteine, and mercapturic acid conjugates) of many foreign compounds, are so efficiently secreted in the bile. Usually only mercapturic acid conjugates but sometimes also the cysteine conjugates are eliminated in the urine. The GSH conjugates and their catabolites in the bile may be further metabolized by the intestine and by microorganisms in the intestine to compounds not containing the cysteine moiety. These metabolites may also be reabsorbed, and may undergo further metabolism or enterohepatic circulation, or they may be excreted in the feces. These complicating factors probably explain why mercapturic acid conjugates are not observed more frequently in the urine, milk, eggs, and tissues of animals exposed to foreign compounds. Thus, as HUTSON (1976), CHASSEAUD (1976), and others have pointed out, the amount of mercapturic acids in the urine is not an adequate measure of GSH conjugation. The best way to study GSH conjugation *in vivo* is to analyze the bile collected from surgically modified animals (HUTSON 1975).

A wide variety of compounds is conjugated with GSH (in reactions catalyzed by a spectrum of transferase enzymes), but these substrates all share a unifying characteristic—either they have an electrophilic center or they are metabolized to a compound with an electrophilic center (e.g., by epoxidation) before GSH conjugation (HUTSON 1970, 1972, and 1975, HOLLINGSWORTH 1977, GROVER 1977, BOYLAND 1971, CHASSEAUD 1973 and 1976, WOOD 1970).

As HUTSON (1975) and others have discussed, most GSH conjugation reactions are similar to an SN2 reaction of a sulfur nucleophile with an electrophilic carbon, and this mechanism is the driving force for these reactions. However, there is also evidence that a carbonium ion intermediate may be involved in the formation of GSH conjugates of certain compounds (BOYLAND and CHASSEAUD 1969).

It is difficult to design a classification system for the wide spectrum of compounds that are GSH conjugates (Fig. 8). As will be discussed later, there is growing evidence that some of the classification systems that have been used (HUTSON 1975, BOYLAND 1971, CHASSEAUD 1973, WOOD 1970) will have to be abandoned or at least modified (GROVER 1977, JAKOBY *et al.* 1976 a and b). Nevertheless, the classification system proposed by CHASSEAUD (1973) will be used here because it serves as a convenient framework for discussing the major classes of compounds that are conjugated by GSH.

The most extensively studied GSH conjugating system is the glutathione S-aryl transferase system (EC 2.5.1.13). Substrates for this enzyme system include a wide variety of simple and complex aromatic ring systems, nitrofurans, and triazine compounds that contain halogens or nitro groups (Figs. 8a–8e). A halogen or nitro group is replaced by GSH (analogous to an SN2 aromatic substitution reaction) (CHASSEAUD 1973).

The GSH conjugation of α- and β-unsaturated compounds (Figs. 8f–8k) is catalyzed by the GSH S-alkene transferase system. The reaction involves the addition of the nucleophile GS$^-$ to an activated double bond and can be classified as an activated vinyl reaction (CHASSEAUD 1973). There is good evidence that some olefins are converted to epoxide intermediates before GSH conjugation (HUCKER 1973). Compounds subject to this type of conjugation include acetals, aldehydes, esters, ketones, lac-

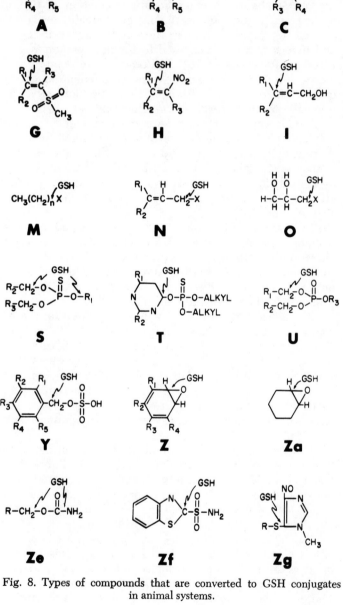

Fig. 8. Types of compounds that are converted to GSH conjugates in animal systems.

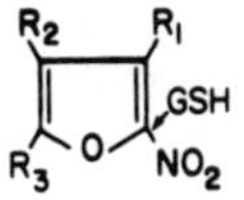

D

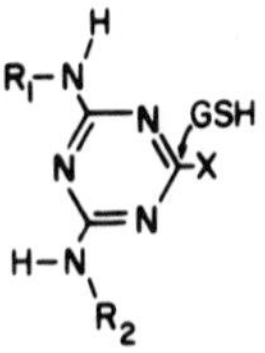

E

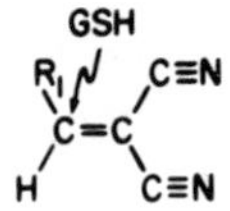

F

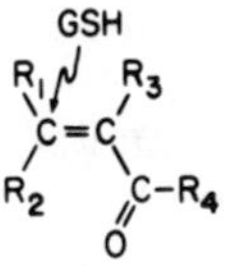

J

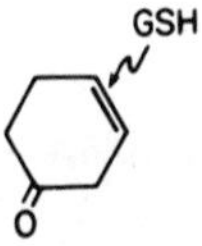

K

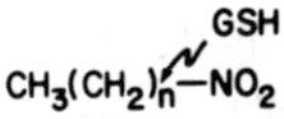

L

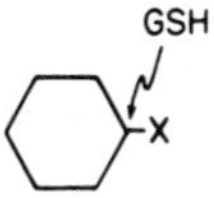

P

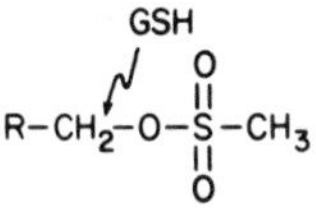

Q

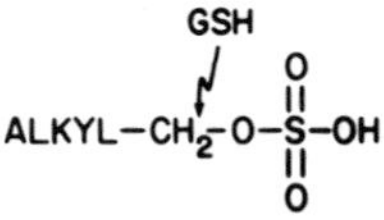

R

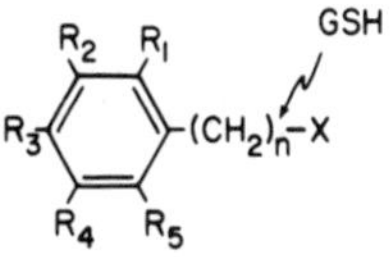

V

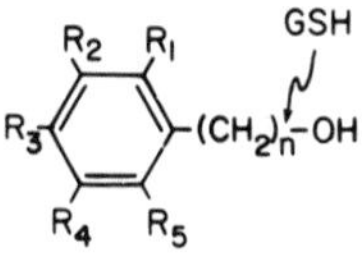

W

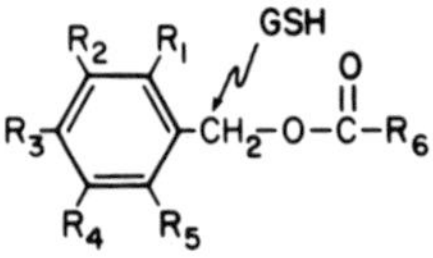

X

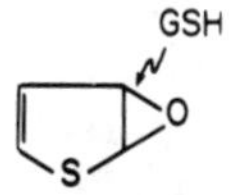

Zb

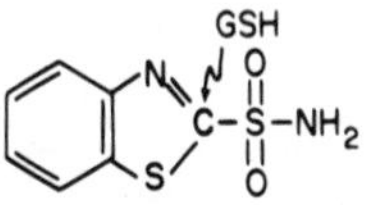

Zc

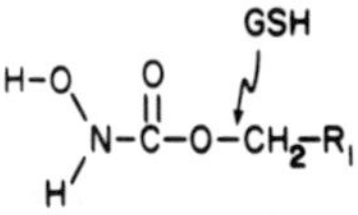

Zd

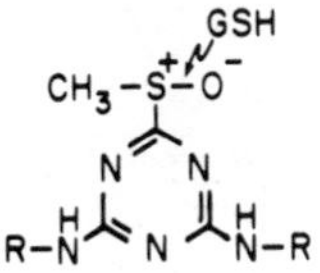

Zh

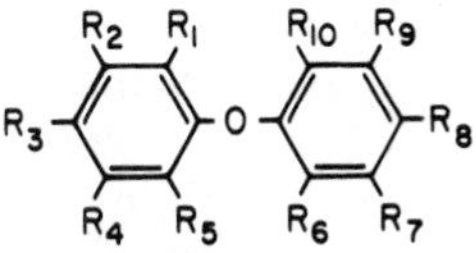

Zi

tones, ntiriles, sulfones, and nitro compounds (BOYLAND 1971, CHASSEAUD 1973 and 1976, WOOD 1970).

The GSH S-alkyl transferase (EC 2.5.1.12) system is responsible for the GSH conjugation of several classes of compounds, including nitroalkanes, alkyl halides, halogenocycloalkanes, alkyl methanesulfonates, and organophosphates (Figs. 8l–8u).

Another transferase system (GSH-arylalkyl transferase; EC 2.5.1.14) catalyzes the GSH conjugation of arylalkyl halides, arylalkyl alcohols, and arylalkyl esters (Figs. 8v–8y). The latter class includes arylalkyl sulfates such as menaphthyl sulfate, that are products of the previously discussed sulfate ester conjugating system.

The involvement of GSH in the metabolism of many simple and polycyclic hydrocarbons is crucially important to the animal and has been the subject of intensive investigation. As previously mentioned, early studies showed that S-(p-bromophenyl) acetylcysteine could be isolated from the urine of animals dosed with bromobenzene. It was recognized that this compound (a mercapturic acid conjugate) was not present in the "free state' but was derived from unknown substances (referred to as "premercapturic acid conjugates") by treatment with acid. Acidification was used routinely in the isolation of products from the urine. It was not until the 1950s that BOYLAND and others (1957) demonstrated that the so-called "premercapturic acid conjugates" were the true metabolites and the "mercapturic acid conjugates" were artifacts of the procedures used to isolate the metabolites of brombenzene and related compounds. These and related studies, which have conclusively demonstrated that epoxide formation is a critical intermediate step in the formation of GSH conjugates of a wide variety of simple and polycyclic hydrocarbons, have been reviewed by WOOD (1970), BOYLAND (1971), CHASSEAUD (1973), and others. The metabolism of naphthalene (Fig. 9) is typical of the metabolism of this class of compounds. The first step is the formation of an epoxide which may then be enzymatically conjugated with GSH. This hydroxy-GSH metabolite may then be converted to the "premercapturic conjugate' 'through the successive losses of glutamate and glycine and the addition of the acetyl group (Figs. 7 and 9). Thus, it is now apparent that the so-called "premercapturic acid conjugate" is the true metabolic product and the mercapturic acid conjugates derived from this class of compounds are artifacts commonly produced by purification techniques (chiefly, acidification).

Figure 9 also demonstrates the crucial role of GSH conjugation in the metabolism of aryl hydrocarbons, many of which are potent carcinogens. The classic studies by MILLER and MILLER and many others have shown that the epoxides (arene oxides) of a number of simple and polycyclic hydrocarbons are the "ultimate carcinogens." The epoxides are electrophiles that have a very high affinity for nucleophilic centers, including those in DNA, RNA, and proteins. These residues, covalently bonded to tissue macromolecules, produce tissue necrosis and in some instances carcinogenesis (HUCKER 1973, MILLER and MILLER 1974). As WOOD

(1970), JERINE and DALY (1974), and others have pointed out, the epoxide may be converted to other compounds both enzymatically and non-enzymatically (Fig. 9). The fate of the epoxide depends on the nature of the rest of the molecule, substrate levels, and other factors. The epoxides may be converted to dihydrodiols by epoxide hydrase and then further metabolized by either oxidation or dehydration to phenols and catechols and ultimately to sulfate ester and glucuronic acid conjugates of these metabolites (WOOD 1970, JERINE and DALY 1974). For many types of compounds, these routes of metabolism (hydration, oxidation, dehydration, etc.) are extremely efficient, and the half-life of the epoxide intermediate is very short (WOOD 1970, JERINE and DALY 1974 and 1976). However, for some types of simple and polycyclic hydrocarbons, the epoxide intermediates are not efficiently metabolized by these routes. In some instances the epoxides react with nucleophilic sites in DNA, RNA, and tissue proteins or they are conjugated with GSH. For such compounds GSH conjugation is especially important as a detoxication mechanism.

In addition to the extensive knowledge of the epoxides of aryl hydro-

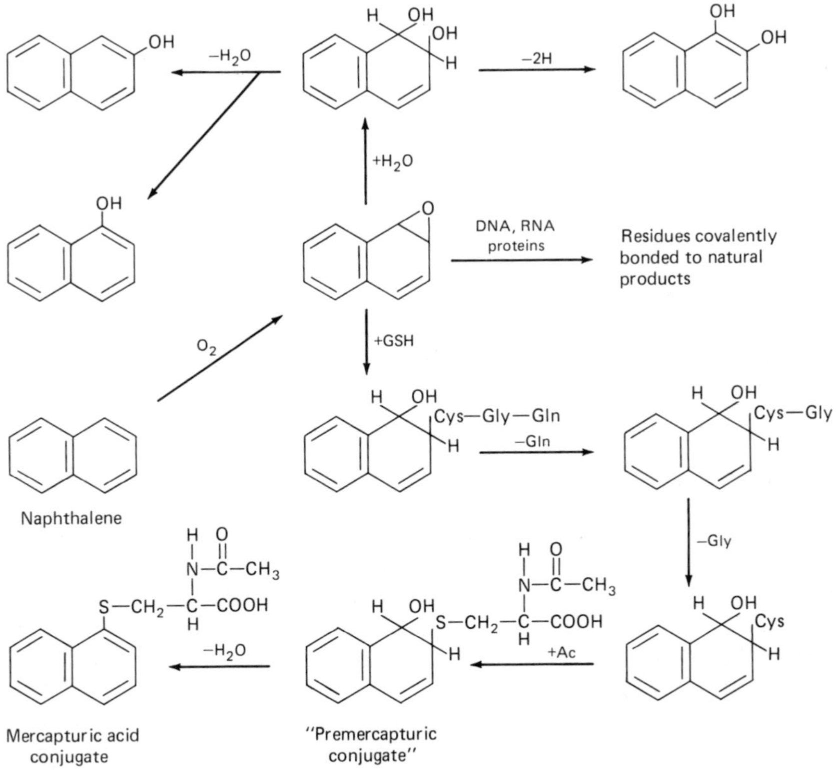

Fig. 9. The involvement of glutathione (GSH) in the metabolism of naphthalene by animal systems.

carbons (Fig. 8z) and aryl halides, it is now known that epoxides of aliphatic compounds (Figs. 8za–8zb) are also conjugated by the GSH S-epoxide transferase system(s) (Hutson 1975, Chasseaud 1973).

Hutson (1975), Chasseaud (1973), Wood (1970), Boyland (1971), and others listed other miscellaneous compounds that are GSH conjugated; these include N-hydroxy compounds, urethanes, bis-β-chloroethyl sulfide, sulfonamides, aromatic amines, and thiophens (Figs. 8zc–8zg). Bedford et al. (1975) reported on the GSH displacement of methylsulphenic acid from a triazine ring (Fig. 8zh). Lamoureux and Davison (1975) studied the formation of a mercapturic conjugate of a diphenyl ether (cleavage of the ether linkage; Fig. 8zi). As the field progresses, it seems apparent that the list of compounds known to be conjugated with GSH will continue to grow.

In recent studies, Jakoby (1978), Jakoby et al. (1976 a and b), were the first to isolate a series of GSH S-transferase enzymes, each as homogeneous proteins (designated AA, A, B, C, D, E, and M). Kinetic and substrate specificity studies with these homogenous enzymes have led Jakoby (1978), Jakoby et al. (1976 a and b) and Grover (1977) to question the validity of the GSH S-transferase classification system proposed by Chasseaud (1973) (and used in this review). It seems apparent that at least some modification of the classification system is imminent (Jakoby 1978, Jakoby et al. 1976 a and b).

Grover (1977) and Jakoby (1978) have summarized the information on several homogeneous GSH S-transferases; it indicates that they all have a molecular weight of 45,000 to 50,000 and are composed of 2 subunits of approximately equal size. Grover (1977) speculated that one subunit of each enzyme moiety served as the binding site for GSH (as the thiolate ion) and that the other subunit had a less specific binding site for electrophilic substrates. He suggested that the glutathione transferases may be a family of enzymes, the molecules of which consist of a common subunit that binds GSH and any one of several proteins that have a binding site for electrophilic substrates (Grover 1977).

Recent studies demonstrated the important relationship between ligandin and a GSH S-transferase (Grover 1977, Jakoby 1978). The terms "ligandin" and "basic azo-dye binding protein" have been used for several years to describe a soluble hepatic protein that binds many compounds, including bilirubin, some steroids, cholecystographic agents, azo dyes, carcinogenic hydrocarbons, and other foreign compounds (Jakoby 1978, Arias et al. 1976, Listowsky et al. 1976). It had been postulated that ligandin serves only as a special "transport protein." However, recent studies have shown that ligandin is identical to a GSH S-transferase (GSH-transferase B) (Jakoby 1978, Jakoby et al. 1976 a and b, Sarrif and Heidelberger 1976). Certain types of natural products and foreign compounds, as indicated above are bound to the "ligandin" protein (sometimes with covalent bonding, thereby destroying the proteins' GSH S-transferase activity), but the compounds are not substrates for GSH conjugation. Such a compound is bound to the enzyme even though it has no electrophilic center for reaction with the thiolate ion of GSH or the molecule is held in a configuration, which does not allow the reaction to proceed. Thus, there is evidence that at least one GSH S-transferase enzyme can operate in 3 ways to eliminate foreign compounds. It can (1) catalyze GSH conjugation yielding a product with ideal physicochemical properties for elimination in the bile, (2) form a noncovalent bond thereby serving as a transport protein carrying the molecule to a site of metabolism or excretion, or (3) form a covalent bond inactivating both the compound and the enzyme (GROVER 1977, JAKOBY 1978). Further research is required to determine the relative quantitative importance of these 3 mechanisms in different animals and tissue; however, the information currently available indicates that they are all important in scavenging potentially harmful compounds from the cell.

Other sulfhydryl-containing compounds, such as cysteine, cysteine peptides, and proteins, apparently react with activated hydrocarbons *in vitro*, but such reactions are apparently much slower than GSH conjugation. The GSH concentration in mammalian tissue represents by far the largest portion of the available sulfhydryl-containing compounds and only the reaction with GSH is known to be enzyme catalyzed (WOOD 1970). There is evidence that an S-substituted cysteine moiety in protein may be released when the protein is degraded *in vivo* and then acetylated and excreted in the urine as the mercapturic acid conjugate (WOOD 1970). The quantitative importance of this pathway probably varies greatly from compound to compound (see previous discussion), but it seems likely that it would be most prevalent when tissue GSH levels are low or when the GSH-conjugating mechanism is impaired. There is persuasive evidence that tissue levels of GSH levels can be depleted by the administration of foreign compounds and that such depletion profoundly increases the toxicity of at least some classes of foreign compounds (HUTSON 1975, JONES 1973, MITCHELL *et al.* 1976).

Although GSH S-transferase enzymes are normally present in large amounts (about 10% of the soluble liver proteins), the synthesis of these enzymes is induced by exposure of animals to certain foreign compounds (GROVER 1977, KAPLOWITZ and CLIFTON 1976).

There is also evidence for mixed multiple conjugation involving GSH. For instance, JAVITT (1976) reported that a metabolite of tetrabromophenolphthalein contained both GSH and a glucuronic acid moiety.

BARNSLEY (1964), SKLAN and BARNSLEY (1968), and others have reported that S-methylcysteine is metabolized to a variety of compounds including methylmercapturic acid, methylthioacetic acid, methylmercapturic acid sulfoxide, methylthioacetic acid sulfoxide, methylsulfinylacetic acid and 2-hydroxy-3-methyl-sulfinylpropionic acid, N-(methylthioacetyl) glycine, and inorganic sulfate. As WOOD (1970) pointed out, S-methylcysteine may be a special case, but there is ample evidence that at least some other mercapturic acid conjugates may also be further metabolized to phenols, thiophenols, deacetylated compounds, and inorganic sulfate (JONES 1973). COLUCCI and BUYSKE (1965) observed that benzothiazole-2-glutathione (or its cysteine derivatives) was cleaved at the sulfur-carbon bond by a "thionase" to yield benzothiazole-2-mercaptan. The latter was then converted to benzothiazole-2-mercaptoglucuronide. It was demonstrated that the sulfur atom at the 2-position was from GSH and that the reaction occurred both *in vivo* (in the rat, rabbit, and dog) and *in vitro* (in mitochondria and soluble fractions). ANDERSON and SCHULTZE (1965) reported the presence in bovine liver and kidney of a C-S lyase that cleaved S-(1,2-dichlorovinyl)-L-cysteine.

VIII. Methylation

Although methylation is generally not as important quantitatively as other types of conjugation and has not been studied as extensively, there is ample evidence that a wide variety of foreign compounds is methylated by animal systems (WILLIAMS 1967, HUTSON 1972, AXELROD 1962, 1971, MUDD 1973, WILLIAMS 1971).

MUDD (1973) reviewed the original studies by HOFMEISTER, CHALLENGER, RIESER, DU VIGNEAUD, and others who provided the first demonstration of methylation of endogenous compounds. He also summarized studies in the 1940s and early 1950s that led to the demonstration that the sulfonium compound S-adenosyl methionine (SAM) serves as the source of the methyl group MUDD 1973). The methyl carbon is readied for nucleophilic attack by electron withdrawal to the sulfonium group. The nucleophilic substrate (led by N, S, or O) bonds with the methyl moiety and displaces the S-adenosyl homocysteine molecule. In this way, the methyl group of SAM may be transferred to the nitrogen of primary, secondary, or tertiary amines; certain nitrogen containing heterocyclic compounds; the oxygen of phenols and certain N-hydroxy compounds (WILLIAMS 1967, AXELROD 1971, MUDD 1973, WILLIAMS 1971, BREMER and GREENBERG 1961, GESSNER and JAKUBOWSKI 1972, LOTLIKAR 1968), and the sulfur of thiols and certain related compounds (Fig. 10).

Methionine adenosyltransferase (EC 2.5.1.6), the enzyme that catalyzes the formation of SAM from ATP and methionine (Fig. 11), is widely distributed in animal systems. The mechanism, substrate specificity, kinetics, and regulation of this enzyme have been reviewed in detail (MUDD 1973). There is apparently no other mechanism for the biosynthesis of SAM in biological systems (MUDD 1973).

The methylation reaction most extensively studied is the one catalyzed by catechol-O-methyltransferase (EC 2.1.1.6). This enzyme has been investigated primarily with endogenous substrates (USDIN and SNYDER 1973); however, structurally related foreign compounds are methylated by the same system (KATZ and JACOBSON 1973, CRUEUELING et al. 1972). Methylation by this enzyme occurs only when the substrate molecule contains 2 or more adjacent hydroxyl groups. Either hydroxyl group may be methylated by the enzyme, but not both; the position that is favored

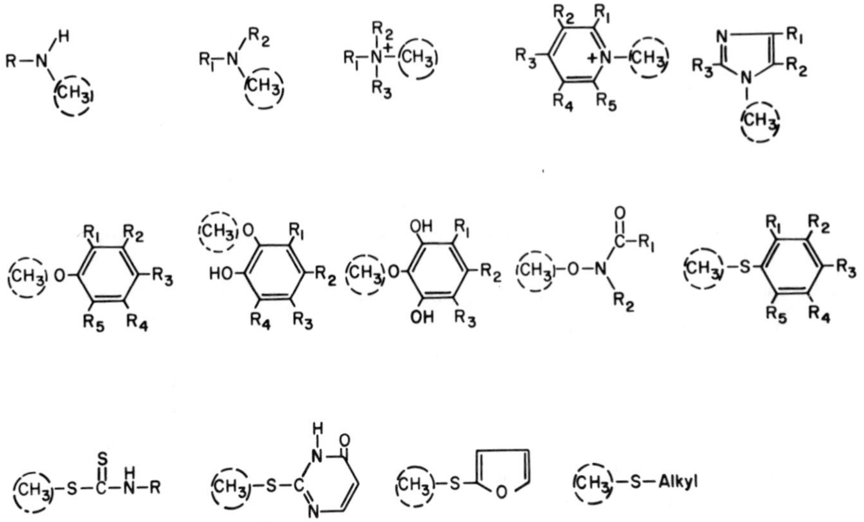

Fig. 10. Types of methylated foreign compounds formed in animal systems.

varies and depends upon the other substitutions on the ring (WILLIAMS 1967, HOLLINGSWORTH 1977, HUTSON 1975, AXELROD 1971, FRÈRE and VERLY 1971, IQBAL and MENZER 1972, McBAIN and MENN 1969, REMY 1963, LADURON et al. 1974). The apparent interconversion of hydroxy-methoxy compounds in vivo results from enzymatic demethylation followed by methylation of the other hydroxyl group (AXELROD 1971). Thus, the relative amounts of para- and meta-methylated compounds eliminated in the urine may be affected by the amount of demethylation that occurs. The p-methoxy group is more rapidly cleaved than the m-methoxy group by hepatic demethylase (WILLIAMS 1967).

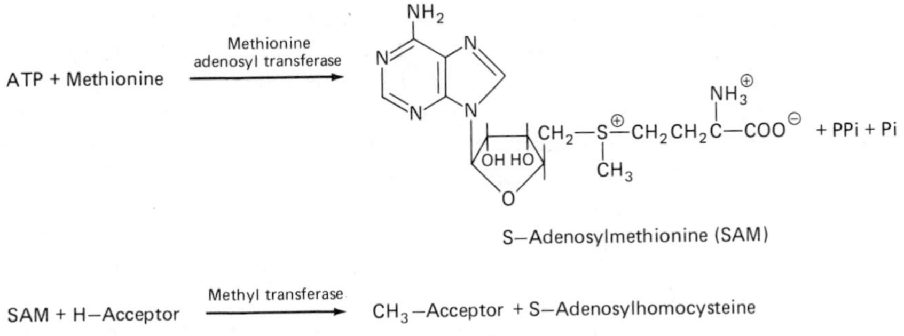

Fig. 11. The biosynthesis of S-adenosylmethionine (SAM) and methylated foreign compounds in animal systems.

When the compound contains 3 adjacent hydroxyl groups (e.g., pyrogallol), the center hydroxyl group is methylated in rats and rabbits by catechol-O-methyltransferase (WILLIAMS 1967). However, if one of the "outside" hydroxyl groups is converted to a methoxy group, either (but not both) of the 2 remaining hydroxyl groups may be methylated (WILLIAMS 1967, AXELROD 1971).

There is evidence for at least 2 forms (isoenzymes) of catechol-O-methyltransferase in hepatic tissue from man, dog, and cat; the activity is present primarily in the cytosol, but there is evidence for a small amount of activity in the microsome fraction (AXELROD 1971, FRÈRE and VERLY 1971).

Catechol-O-methyltransferase has been observed in a wide variety of mammalian tissues, including the liver, kidney, spleen, intestines, salivary gland, pituitary, thyroid, pineal body, aorta, heart, central nervous system, and skin (HARTIALA 1973, AXELROD 1971). Its activity has been measured in many animals, including the rat, cow, pig, mouse, guinea pig, cat, rabbit, and man (WILLIAMS 1967).

Phenol O-methyltransferase (EC 2.1.1.25) catalyzes the O-methylation of a wide variety of alkylphenols, methoxyphenols, and halophenols. This enzyme is localized in the microsome fraction and has been observed in the liver, lung, and other tissues of mammals (AXELROD 1971).

Hydroxy indole-O-methyltransferase, an enzyme localized in the pineal body, catalyzes the methylation of N-acetylserotonin to form melatonin. N-Acetylserotonin is the preferred substrate for this enzyme, although a few closely related compounds are also methylated by it (AXELROD 1971). Apparently, whether this enzyme may be involved in the metabolism of foreign compounds is unknown.

AXELROD (1971) reviewed the evidence for "non-specific N-methyltransferase," which N-methylates a wide variety of endogenous and foreign compounds (both primary and secondary amines). This nonspecific N-methylase is in the cytosol and has been observed in the lung, liver, spleen, adrenal gland, and brain of the chicken, rat, rabbit, mouse, guinea pig, and cat (WILLIAMS 1967, AXELROD 1971).

Phenylethanolamine N-methyl transferase catalyzes the SAM-mediated N-methylation of norepinephrine to epinephrine and the N-methylation of other endogenous compounds and a variety of foreign compounds. Secondary amines may also be methylated by this enzyme, but they are, in general, poorer substrates than primary amines (AXELROD 1971). There is evidence for at least 5 separate forms of this enzyme in mammalian systems, and activity has been observed in the heart, kidney, brain, and adrenal glands (AXELROD 1971).

A soluble SAM dependent enzyme catalyzes the methylation of the N-hydroxy group in N-hydroxy-2-acetyl aminofluorene, and the methylated product is more carcinogenic than the parent compound (AXELROD 1971). Related N-hydroxy compounds also serve as substrates for this SAM-dependent methyl transferase enzyme (IRVING 1970, AXELROD 1971).

A wide range of exogenous sulfhydryl compounds are methylated by the SAM-requiring microsomal enzyme thiol methyltransferase (EC 2.1.1.9), which is present in a number of animal systems (WILLIAMS 1967 and 1971, HUTSON 1972, AXELROD 1971, MUDD 1973, BREMER and GREEN-BERG 1961, GESSNER and JAKUBOWSKI 1972, IQBAL and MENZER 1972). REMY (1963) reported that the cytosol of mammalian tissues contained another SAM-specific transmethylase which catalyzed the S-methylation of certain pyrimidines and thiopurines. The resulting S-methylated compounds may be further metabolized by the animal before excretion (GESSNER and JAKUBOWSKI 1972, IQBAL and MENZER 1972, McBAIN and MENN 1969).

All the methylating systems discussed to this point utilize SAM as the methyl donor; however, LADURON et al. (1974) reported that 5-methyl tetrahydrofolic acid (5-MTHF) served as the methyl donor in the N-methylation of biogenic amines by rat brain. Whether 5-MTHF or other potential one-carbon donors may be important in the methylation of a variety of foreign compounds apparently has not been investigated in detail.

IX. Acylation

Mammalian systems have the ability to acetylate the NH$_2$, OH, and SH groups of certain endogenous compounds (e.g., choline, coenzyme A, amino sugar). However, the acetylation of foreign compounds in animal systems is restricted primarily, if not exclusively, to those compounds containing an NH$_2$ group (aromatic and aliphatic amines, amino acids, hydrazines, hydrazides, and sulfonamides; Figs. 12 and 13). Apparently, secondary amines and amides are not acetylated, and the diacetyl derivatives [R-N(COCH$_3$)$_2$] of primary amines are not produced by animal metabolism (WILLIAMS 1967). Many studies, especially of aryl amines and sulfonamide drugs, have established that acetyl CoA is the acetyl donor in these reactions (WILLIAMS 1967; HUTSON 1970, 1972, and 1975; WEBER 1971 and 1973). Description of the biosynthesis of acetyl CoA and the importance of this compound in intermediary animal metabolism are available in any current biochemistry textbook.

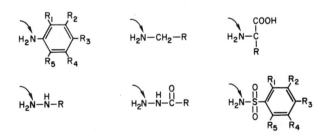

Fig. 12. Types of foreign compounds acylated by animal systems.

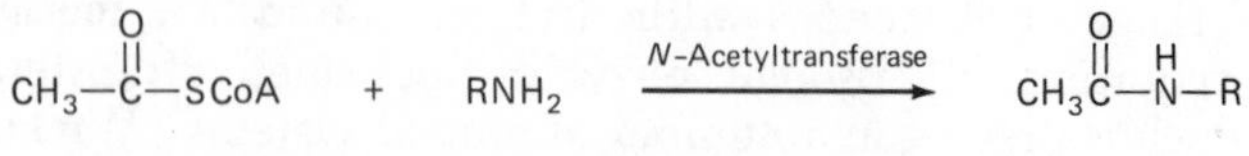

$$\underset{\text{CH}_3-\overset{\displaystyle\overset{\text{O}}{\|}}{\text{C}}-\text{SCoA}}{}\quad+\quad\text{RNH}_2\quad\xrightarrow{\text{\textit{N}-Acetyltransferase}}\quad\underset{\text{CH}_3\overset{\displaystyle\overset{\text{O}}{\|}}{\text{C}}-\overset{\text{H}}{\text{N}}-\text{R}}{}$$

Fig. 13. The acetylation of foreign compounds in animal systems.

The enzyme aryl amine N-transferase (EC 2.3.1.5) is located in the cytosol and is broadly specific for acetyl acceptors (*i.e.*, numerous aryl amines and sulfonamides are acetylated). This enzyme has been observed in a wide variety of tissues, including the liver, gut, lung, thymus, ovary, spleen, uterus, adrenals, leukocytes, kidney, bone marrow, salivary gland, pancreas, pineal body, erythrocytes, brain, muscle and kidney; and in many animals, including the cat, cow, horse, sheep, goat, monkey, mouse, rabbit, rat, guinea pig, and man (WILLIAMS 1967, WEBER 1971 and 1973). However, the capacity to acetylate individual compounds varies greatly with the species and age of the animal (WEBER 1971 and 1973). Studies exhaustively reviewed by WEBER (1971 and 1973) clearly demonstrated that (1) closely related compounds may be acetylated to different degrees by animals of the same species, and (2) animals of the same species may have very different acetylating abilities. Comprehensive studies demonstrated that humans and rabbits are either "rapid acetylators" or "slow acetylators" and that the relative ability to acetylate foreign compounds is inherited as a simple Mendelian trait (WEBER 1971 and 1973). These important discoveries resulted from the clinical observation that normally therapeutic doses of isoniazid were toxic to certain people ("slow acetylators").

There is considerable evidence that more than one N-acetyltransferase enzyme is involved in the acetylation of aryl amines and related compounds. Some data suggest that these enzymes have different substrate specificities and that their relative activities vary from tissue to tissue (LADURON *et al.* 1974, WEBER 1971, HEARSE and WEBER 1973). Furthermore, there is evidence that one form (localized primarily in the liver and gut) makes an individual either a "slow" or "fast" acetylator through its genetically controlled polymorphism, and that at least one other form exists that is not under the same genetic control. The latter enzyme(s) appears to be more widely distributed throughout the body tissues. Although some conflicting data exist (GORDON *et al.* 1973), the bulk of the evidence suggests that at least 2 distinct N-acetyltransferases are present in different tissues and that they are present in different proportions. Complicating the picture even further, there is evidence that the relative activities of the different N-acetyltransferase enzymes vary with the age of the animal (WEBER 1973).

It was formerly thought that foreign compounds containing an aliphatic amine were not N-acetylated (WILLIAMS 1967, WEBER 1971 and 1973), but more recent studies revealed an exception. SULLIVAN *et al.* (1973) reported that rats metabolized α-1-methadol to α-1-6-acetamido-4,4-diphenyl-3-heptanol (*i.e.*, acetylated a primary aliphatic amine).

Some workers have concluded that certain hydroxyl amines are N-acetylated *in vitro*; but the evidence has been questioned (IRVING 1970). For instance, it was reported that N-2-fluorenyl hydroxylamine was acetylated *in vivo* to form N-acetyl-N-2-fluorenyl hydroxylamine, and that the latter compound was then conjugated with glucuronic acid. As IRVING (1970) pointed out, however, the N-2-fluorenyl hydroxylamine may have been sequentially reduced, acetylated, N-hydroxylated, and finally conjugated with glucuronic acid.

HUTSON (1970) and WILLIAMS (1971) discussed a few other types of acylation (formylation and succinylation) of foreign compounds by animal systems. To the author's knowledge, such reactions have not been intensively studied; however, at present there is no evidence that these conjugation mechanisms are quantitatively important for a wide variety of foreign compounds.

Although N-acetylation of foreign compounds frequently reduces or destroys the biological activity of these compounds, there are some important exceptions. There is compelling evidence (IRVING 1970 and 1973, MILLER and MILLER 1969, WEISSBURGER *et al.* 1972, DEBAUN *et al.* 1970) that N-acetylation is the first step in the conversion of certain carcinogenic aromatic amines (e.g., 2-aminofluorene, 4-aminobiphenyl, and 2-amino napthalene) to the ultimate carcinogens.

X. Amino acid conjugation

Foreign compounds containing a free carboxyl group are often excreted as an amino acid conjugate by animals. The most common conjugation moiety is glycine. However, as will be discussed later, other amino acids and related compounds are sometimes important.

Aryl and arylalkyl acids (e.g., benzoic and phenylacetic acids) are most commonly conjugated with amino acids; however, other heterocyclic and carbocyclic compounds also undergo this type of conjugation in animal systems (Fig. 14). CONTI and BICKEL (1977) have reviewed in detail the early studies that ultimately led to the isolation and identification of the glycine conjugate of benzoic acid in the urine of animals dosed with the latter compound.

It should be emphasized that the biosynthesis of amino acid conjugates differs strikingly from the biosynthesis of the other types of conjugates. The previously described conjugations (except GSH conjugation) all involve the activation of an endogenous conjugating moiety (e.g., glucuronic acid and sulfate) that serves as the high-energy donor in the transferase reactions. In contrast, in amino acid conjugation the foreign compound is activated, and this high-energy compound serves as the donor and the conjugating moiety (amino acid) serves as the acceptor in the transferase reaction (Fig. 15). The first step is the conversion of the foreign compound to the foreign compound-S-CoA under the influence of the appropriate acyl CoA synthetase (an acid-thiol ligase) and the second step is the transfer of the activated compound to the acceptor.

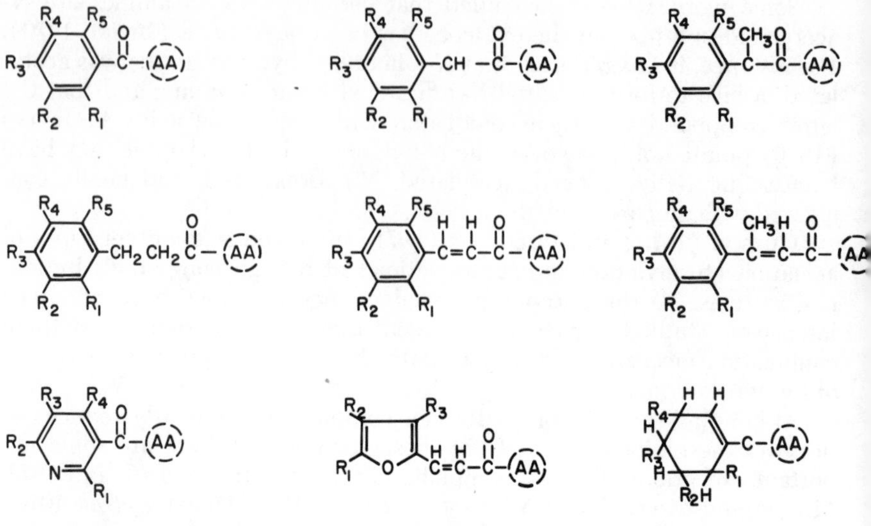

Fig. 14. Types of foreign compounds conjugated with amino acids in animal systems.

Climie and Hutson (1978) have summarized the evidence that benzoic acid is converted to benzoyl-CoA under the influence of the mitochondrial enzyme butyryl-CoA synthetase (EC 6.2.1.2). The transferase enzyme responsible for the conjugation (glycine acyltransferase EC 2.3.1.13) is also in mitochondria (Williams 1967, Climie and Hutson 1978). In contrast, cholyl-CoA synthetase (EC 6.2.1.7), the enzyme responsible for activation of cholate prior to conjugation is a microsomal enzyme (Climie and Hutson 1978). Furthermore, there is evidence (although in some ways it is conflicting) that the transferase(s) enzymes responsible for the glycine and taurine conjugation of cholyl-CoA is present in the soluble, microsomal and lysosomal fractions of animal cells

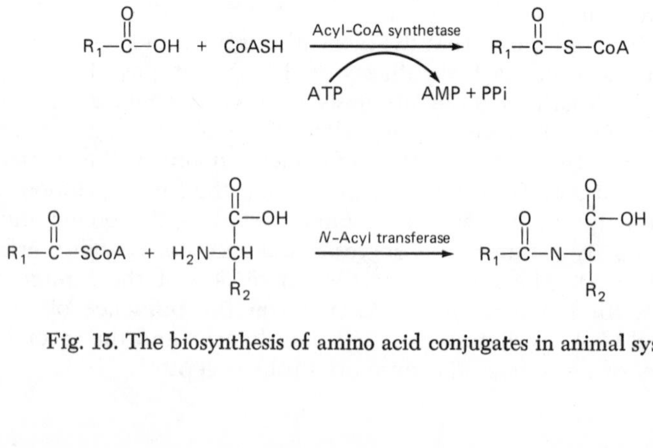

Fig. 15. The biosynthesis of amino acid conjugates in animal systems.

(CLIMIE and HUTSON 1978). Apparently lacking are comprehensive studies to determine whether some foreign compounds are activated and amino acid-conjugated by cell fractions other than the mitochondria. Information on the kinetics, substrate specificities, and other characteristics of the enzymes responsible for the activation and amino acid conjugation of natural products has been summarized by CLIMIE and HUTSON (1978).

There is general agreement that glycine is the most common conjugating moiety; however, glutamine, glutamic acid, serine, alanine, ornithine, arginine, taurine, and glycyltaurine have also been identified as the conjugating moiety in some animal systems (HOLLINGSWORTH 1977, CLIMIE and HUTSON 1978). Species variations are pronounced. For example, glycine conjugation occurs in man, dog, cat, rat, rabbit, mouse, bovine, goat, sheep, pig, monkey, guinea pig, horse, camel, elephant, and some birds but apparently not in domestic chickens, geese, ducks, and turkeys (WILLIAMS 1967 and 1971, HARTIALA 1973, HOLLINGSWORTH 1977). In man and certain monkeys, arginine and glutamine conjugation are also very common. Birds and some reptiles commonly use ornithine as the conjugating moiety (WILLIAMS 1967 and 1971, HARTIALA 1973, HOLLINGSWORTH 1977). Taurine (2-aminoethanesulfonic acid) conjugation of natural products (e.g., bile acids) is very common and at least some foreign compounds are taurine-conjugated by animals (JAMES et al. 1972, GRÜNOW and BOEHME 1974, JORDAN and RANCE 1974, WILLIAMS 1974).

XI. Phosphate and other miscellaneous conjugation reactions

There is evidence that some foreign compounds may be conjugated with phosphate in the animal. LOTIKAR and WASSERMAN (1970) reported that the phosphate ester of 2-(N-hydroxy)acetamidofluorene reacted with methionine and guanosine in vitro. They suggested that this type of conjugate may be an "ultimate carcinogenic metabolite" of aromatic hydroxamic acids in vivo. BOYLAND et al. (1961) reported that dogs dosed with 2-naphthylamine excreted di-(2-amino-1-naphthyl) hydrogen phosphate in the urine; they suggested that the phosphate conjugate is the probable cause of bladder cancer in dogs fed 2-naphthylamine (BOYLAND et al. 1961). Cats dosed with phenol excreted phenyl phosphate in the urine (CAPEL et al. 1974). Although the literature contains these examples of foreign compound phosphorylation, the information currently available suggests that at most this conjugation mechanism is of minor importance in animal systems. This conclusion seems somewhat surprising in light of the qualitative and quantitative importance of phosphorylation of endogenous compounds. Presumably, the wide variety of enzymes involved in the phosphorylation of natural products are highly substrate specific.

WILLIAMS (1967) has discussed a few other types of conjugates that have been reported; these include glycyltaurine, formyl, and peptide conjugates. Certainly, these reports cannot be discounted. However, there is no strong evidence that such conjugations are common in animals.

XII. Factors affecting conjugation reactions

A few examples of factors that affect specific conjugation reactions were cited in previous sections. An in-depth discussion of all the factors affecting conjugation reactions in animal systems is beyond the scope of this review, but a few generalizations and comments on the more important factors will be made. More complete discussions are available in a number of papers and reviews (FISHMAN 1970; LEE et al. 1977; WILLIAMS 1947, 1959, 1967, and 1974; HATHWAY 1970, 1972, and 1975; LADU et al. 1971; PARKE 1968; BRODIE and GILLETTE 1971; PARK and SMITH 1977; DUTTON et al. 1971 and 1977; SMITH and WILLIAMS 1966; DUTTON 1966; MIETTINEN and LESKINEN 1970; DUTTON and BURCHELL 1977; HUTSON 1970, 1972 and 1975; MILLBURN 1974; HUNTER and CHASSEAUD 1976; DODGSON and ROSE 1970; PAULSON 1976; ROY 1960 and 1971; GREGORY and ROBBINS 1960; ROBBINS 1962; DEMEIO 1975; PECK 1974; HOLLINGSWORTH 1977; GREGORY 1962; BOYLAND and CHASSEAUD 1969; CHASSEAUD 1973).

The age of the animal is generally recognized to be of critical importance (WILLIAMS 1967; HATHWAY 1970, 1972, and 1975; LADU et al. 1971; PARKE 1968; BRODIE and GILLETTE 1971; PARKE and SMITH 1977; DUTTON 1966 and 1971; HUTSON 1970; SMITH 1968; HOLLINGSWORTH 1977). With certain notable exceptions, tissues from fetuses and from very young animals and very old animals conjugate foreign compounds less efficiently than tissues from adolescent and adult animals.

Studies, both in vitro and in vivo, have conclusively demonstrated that the rate of conjugation by male and female animals of the same species may differ strikingly (WILLIAMS 1967; HATHWAY 1970, 1972, and 1975; LADU et al. 1971; PARKE 1968; BRODIE and GILLETTE 1971; PARKE and SMITH 1977; DUTTON 1966 and 1971; KENNEDY et al. 1978; HUTSON 1970; SMITH 1968; HOLLINGSWORTH 1977; KING and OLIVE 1975; CHABRA and TOUTS 1974; OEHME 1970; MOORE 1972). These differences may be quantitative, qualitative (via different pathway), or both.

It is also well documented that the conjugation of a foreign compound in one species may be (and very often is) completely different from the conjugation of the same compound in another species (WILLIAMS 1967 and 1974; HATHWAY 1970, 1972, and 1975; LADU et al. 1971; PARKE 1968; BRODIE and GILLETTE 1971; PARKE and SMITH 1977; DUTTON 1966 and 1971; HUTSON 1970; SMITH 1968; HOLLINGSWORTH 1977; KING and OLIVE 1975; OEHME 1970; MOORE 1972; BRODIE and MAICKEL 1962). Thus, generalizations about the metabolism of a compound that are based on studies with only one species are at best risky and are generally not acceptable. The literature abounds with examples of compounds that are very deleterious (e.g., acutely or chronically toxic, carcinogenic) to one species but which have little or no effect on another species. Often the differences in biological activity of the compound can be directly attributed to qualitative or quantitative differences in the conjugation of the compound by the different species.

Certain conjugation mechanisms are impaired or completely absent in some animals, even though others of the same species have full capability (e.g., Gunn rat, slow acetylating rats and human beings) (WILLIAMS 1967; HATHWAY 1970, 1972, and 1975; LADU et al. 1971; PARKE 1968; BRODIE and GILLETTE 1971; PARKE and SMITH 1977; DUTTON 1966 and 1971; MIETTINEN and LESKINEN 1970; HUTSON 1970; SMITH 1968; HOLLINGSWORTH 1977; KING and OLIVE 1975; WEBER 1973; GORDON et al. 1973; MOORE 1972).

Size of dose (HATHWAY 1970, 1972, and 1975; LADU et al. 1971; DUTTON 1966; SMITH 1968; HOLLINGSWORTH 1977) and route of administration (WILLIAMS et al. 1974, CONWAY et al. 1973, LEVY 1968 and 1971, LEVY et al. 1972) may markedly effect the conjugation of a foreign compound. HARTIALA (1973 and 1977) and others have reviewed the wide variety of biotransformations that can occur in the gut. Thus, it is not surprising that a compound given orally may be conjugated differently than the same compounds given by injection. The differing rates and routes of metabolism that occur after administration of different size doses is generally thought to be due to depletion of conjugating substrates in the tissues or by exceeding tissue first-order substrate concentrations. Because of the relatively small endogenous sulfate pool, the mechanism for sulfate ester biosynthesis is apparently quite susceptible to saturation by the administration of a large dose of a foreign compound.

Dietary factors (e.g., low levels of sulfur) may markedly affect the relative importance of 2 or more competing conjugation mechanisms (WILLIAMS 1967; HATHWAY 1970, 1972, and 1975; LADU et al. 1967; SMITH 1968; SLOTKIN and DISTEFANO 1970; KURYZNSKE and SMITH 1975; HOLLINGSWORTH 1977; BLUNCK and CROWTHER 1975). This phenomenon is often related to the depletion of conjugating substrate pools.

The enzymes responsible for the conjugation of at least most foreign compounds can be induced (WILLIAMS 1967; HATHWAY 1970, 1972, and 1975; LADU et al. 1971; DUTTON 1966 and 1971; SMITH 1968; HOLLINGSWORTH 1977; HUTSON 1972; GROVER 1977; KAPLOWITZ and CLIFTON 1976; VANIO and AITIO 1974). Therefore, pretreating an animal with a foreign compound may profoundly affect the rates and routes of conjugation of that compound and, in many cases, other compounds that may be conjugated by the same system(s).

It is not surprising that a number of other factors such as disease states (WILLIAMS 1967; HATHWAY 1970, 1972, and 1975; LADU et al. 1971; SMITH 1968; GESSNER 1974; HOLLINGSWORTH 1977; BLUNCK and CROWTHER 1975), and environmental temperature (WILLIAMS 1967, PARKE 1968, SMITH 1968, HOLLINGSWORTH 1977) may also affect conjugation. Moreover, the possibilities are almost unlimited for the interaction of 2 or more of these factors to affect the rates and routes of conjugation in the animal system. It seems inevitable that, as research in this field continues, additional factors (e.g., diet, stress, and inherited conditions) will be recognized as having an effect on conjugation reactions.

The implications of the susceptability of conjugation reactions to such a wide range of factors and interactions are so important that they cannot be ignored. Especially important is an awareness of these factors as they relate to the administration of therapeutic doses of drugs to animals and man, and to the use of agricultural chemicals in the environment.

Summary

Animal cells have enzyme systems in strategic locations (cytosol, mitochondria, and endoplasmic reticulum) to conjugate a wide variety of foreign compounds that contain either electrophilic or nucleophilic centers. Animal systems form (1) glucuronic acid and related carbohydrate conjugates, (2) sulfate ester conjugates, (3) glutathione and related mercapturic acid conjugates, (4) methylated products, (5) acylated products, (6) amino acid and related conjugates, and (7) a few other (much less important quantitatively) miscellaneous conjugates. The relative qualitative and quantitative importance of these conjugation reactions is dependent upon many factors which include (1) the nature of the molecule; (2) the species, sex, age, and physiological and nutritional state of the animal; and (3) the size and route of administration of the dose. All of the conjugating systems are energetically expensive anabolic processes involving the biosynthesis of high energy nucleotides (or utilize a high energy equivalent to generate the appropriate intermediate) that serve as substrates for the conjugating transferase enzymes. The conjugates are in general more polar and therefore usually more efficiently secreted in the urine and/or excreted in the bile than the parent compound. These and other observations during the mid to late 19th century led to the conclusion that the conversion of foreign compounds to polar conjugates is always a detoxication process. Although this is undoubtedly true for many (probably the majority) of the foreign compounds that animals are exposed to, there are notable exceptions. More recent studies have shown that in some instances, conjugation may result in a product with greater biological activity than the parent compound.

References

Aitio, A.: Glucuronide synthesis in the rat and guinea pig lung. Xenobiotica. 3, 13 (1973).

Akitake, H., and K. Kobayashi: Studies on the metabolism of chlorophenols in fish— III. Isolation and identification of a conjugated PCP excreted by goldfish. Bull. Japan. Soc. Scientific Fisheries 41, 321 (1975).

Anderson, P. M., and M. O. Schultze: Cleavage of S-(1,2-dichlorovinyl)-L-cysteine by an enzyme of bovine origin. Arch. Biochem. Biophys. 111, 593 (1965).

Arias, M., G. Fleischer, R. Kirsch, S. Mishkin, and Z. Gatmaitan: On the structure regulation and function of ligandin. In I. M. Arias and W. B. Jakoby (eds.): Glutathione: Metabolism and function, p. 175. New York: Raven Press (1976).

Artz, N. E., and E. M. Osman: Biochemistry of glucuronic acid. New York: Academic Press (1950).

Axelrod, J.: Demethylation and methylation of drugs and physiologically active

compounds. In B. B. Brodie and E. G. Erdös (eds.): Metabolic factors controlling
duration of drug action, vol. 6, p. 97. New York: Pergamon Press (1962).
—— Methyl transferase enzymes in the metabolism of physiologically active com-
pounds and drugs. In B. B. Brodie and J. R. Gillette (eds.): Concepts in bio-
chemical pharmacology, part II, p. 601. New York: Springer-Verlag (1971).
BARNSLEY, E. A.: The metabolism of S-methyl-L-cysteine in the rat. Biochem. Biophys.
Acta 90, 24 (1964).
BAUMAN, E.: Ueber das Vorkommen von Brenzcatechin im Harn. Arch. Ges. Physiol.
Pfleugers 12, 63 (1876).
BEDFORD, C. T., M. J. CRAWFORD, and D. H. HUTSON: Sulphoxidation of cyanatryn,
a mercapto-sym-triazine herbicide, by rat liver microsomes. Chemosphere 5,
311 (1975).
BLUNCK, J. M., and C. E. CROWTHER: Enhancement of azo dye carcinogenesis by
dietary sodium sulphate. Eur. J. Cancer 11, 23 (1975).
BOYLAND, E.: Mercapturic acid conjugation. In B. B. Brodie and E. G. Erdös (eds.):
Metabolic factors controlling duration of drug action, Vol. 6, p. 65. New York:
Pergamon Press (1962).
BOYLAND, E.: Mercapturic acid conjugation. In B. B. Brodie and J. R. Gillette (eds.):
Concepts in biochemical pharmacology, part II, p. 584. New York: Springer-
Verlag (1971).
BOYLAND, E., and L. F. CHASSEAUD: The role of glutathione and glutathione S-trans-
ferases in mercapturic acid biosynthesis. Adv. Enzymol. 32, 173 (1969).
BOYLAND, E., C. H. KINDER, and D. MANSON: The biochemistry of aromatic amines.
8. Synthesis and detection of di-(2-amino-1-naphthyl) hydrogen phosphate, a
metabolite of 2-naphthylamine in dogs. Biochem. J. 78, 175 (1961).
——, D. MANSON, and S. F. D. ORR: The biochemistry of aromatic amines. 2. The
conversion of arylamines into aryl sulphamic acids and arylamine-N-glucosiduronic
acids. Biochem. J. 65, 417 (1957).
BREMER, J., and D. M. GREENBERG: Enzymatic methylation of foreign sulfhydryl
compounds. Biochim. Biophys. Acta 46, 217 (1961).
BRODIE, B. B., and J. R. GILLETTE: Handbook of experimental pharmacology, concepts in biochemical pharmacology. II. New York: Springer-Verlag (1971).
——, and R. P. MAICKEL: Comparative biochemistry of drug metabolism. In B. B.
Brodie and G. Erdös (eds.): Metabolic factors controlling duration of drug action,
vol. 6, p. 299. New York: Pergamon Press (1962).
CAPEL, I. D., P. MILLBURN, and R. T. WILLIAMS: Monophenyl phosphate, a new conjugate of phenol in the cat. Biochem. Soc. Trans. 2, 305 (1974).
CARROLL, J., and B. SPENCER: Sulphate activation and sulphotransferases in foetal
and adult rats. Biochem. J. 94, 20 (1965).
CHABRA, R. S., and J. R. TOUTS: Sex differences in the metabolism of xenobiotics by
extrahepatic tissue in rats. Drug Metab. Dispos. 2, 375 (1974).
CHASSEAUD, L. F.: The nature and distribution of enzymes catalyzing the conjugation
of glutathione with foreign compounds. Drug Metab. Rev. 2, 185 (1973).
—— Conjugation with glutathione and mercapturic acid excretion: In M. Arias and
W. B. Jakoby (eds.): Glutathione: Metabolism and function, p. 77. New York:
Raven Press (1976 a).
—— Properties of the glutathione S-alkene transferase system catalyzing the conjuga-
tion of glutathione with diethyl maleate. In I. M. Arias and W. B. Jakoby (eds.):
Glutathione: Metabolism and function, p. 281. New York: Raven Press (1976 b).
CLIMIE, I. J. G., and D. H. HUTSON: Proc. 4th Internat. Mtg., Pest. Chem., Internat.
Union Pure and Applied Chem. Zurich: Pergamon Press (1978).
COLUCCI, D. F., and D. A. BUYSKE: The biotransformation of a sulfonamide to a
mercaptan and to mercapturic acid and glucuronide conjugates. Biochem. Pharma-
col. 14, 457 (1965).
CONTI, A., and M. H. BICKEL: History of drug metabolism: Discoveries of the major
pathways in the 19th century. Drug Metab. Rev. 6, 1 (1977).
CONWAY, W. D., S. M. SINGHUI, M. GIBALDI, and R. N. BOYES: The effect of route

of administration on the metabolic rate of terbutaline in the rat. Xenobiotica 3, 813 (1973).

CRUELING, C. R., N. MORRIS, H. SHIMZUE, H. H. ONG, and J. DALY: Catechol O-methyltransferase. IV. Factors affecting m- and p-methylation of substituted catechols. Mol. Pharmacol. 8, 398 (1972).

DAVIES, D. S.: Drug metabolism in man. In D. V. Parke and R. L. Smith (eds.): Drug metabolism—From microbe to man, p. 357. London: Taylor & Francis Ltd. (1977).

DEBAUN, J. R., E. C. MILLER, and J. A. MILLER: N-Hydroxy-2-acetylamino-fluorene sulfotransferase: Its probable role in carcinogenesis and in protein—(methion-S-yl) binding in rat liver. Cancer Res. 30, 577 (1970).

DEBAUN, J. R., J. Y. ROWLEY, E. C. MILLER, and J. A. MILLER: Sulfotransferase activation of N-hydroxy-2-acetylaminofluorene in the rat. Proc. Amer. Assoc. Cancer Res. 8, 12 (1967).

DEMEIO, R. H.: Phenol conjugation. III. The type of conjugation in different species. Arch. Biochem. 7, 323 (1945).

—— Sulfate activation and transfer in metabolic pathways. In D. M. Greenberg (ed.): Metabolism of sulfur compounds, vol. 7, 3rd ed., p. 287. New York: Academic Press (1975).

DIETERLE, W., J. W. FAIGLE, F. FRÜH, H. MORY, W. THEOBALD, K. O. ALT, and W. J. RICHTER: Metabolism of phenylbutazone in man. Arzneim. Forsch. 26, 572 (1976).

DODGSON, K. S., and F. A. ROSE: Sulfoconjugation and sulfohydrolysis. In E. H. Fishman (ed.): Metabolic conjugation and metabolic hydrolysis, vol. 1, p. 239. New York: Academic Press (1970).

DOROUGH, H. W.: Biological activity of pesticide conjugates. In D. D. Kaufman, G. G. Still, G. D. Paulson, and S. K. Bandal (eds.): Bound and conjugated pesticide residues, p. 11. Washington, D. C.: Amer. Chem. Soc. Symposium Series (1976).

DUGGAN, D. E., J. J. BALDWIN, B. H. ARISON, and R. E. RHODES: N-Glucoside formation as a detoxification mechanisms in mammals. J. Pharmacol. Exp. Ther. 190, 563 (1974).

DUTTON, G. J.: Glucuronic acid, free and combined. New York: Academic Press (1966).

—— Glucuronide forming enzymes. In B. B. Brodie and J. R. Gillette (eds.): Concepts in biochemical pharmacology, vol. 28, part 2, p. 378. New York: Springer-Verlag (1971).

——, and B. BURCHELL: Newer aspects of glucuronidation. In J. W. Bridges and L. F. Chasseaud (eds.): Progress in drug metabolism, vol. 2, p. 1. New York: Wiley (1977).

——, G. J. WISHART, J. E. A. LEAKEY, and M. A. GOHER: Conjugation of glucuronic acid and other sugars. In D. E. Parke and R. L. Smith (eds.): Drug metabolism—From microbe to man, p. 71. London: Taylor & Francis Ltd. (1977).

FISHMAN, W. H.: Metabolic conjugation and metabolic hydrolysis, vols. 1 and 2. New York: Academic Press (1970).

FRÉRE, J. M., and W. G. VERLY: Catechol O-methyltransferase. The para- and meta-O-methylations of noradrenaline. Biochim. Biophys. Acta 235, 73 (1971).

GESSNER, T.: Studies of glucuronidation and sulfation in tumor bearing rats. Biochem. Pharmacol. 23, 1809 (1974).

——, and N. HAMANDA: Identification of p-nitrophenyl glucoside as a urinary metabolite. J. Pharm. Sci. 59, 1528 (1970).

——, A. JACKNOWITZ, and C. A. VOLLMER: Studies of mammalian glucoside conjugation. Biochem. J. 132, 249 (1973).

——, M. JAKUBOWSKI: Diethyldithiocarbamic acid methyl ester; a metabolite of disulfiram. Biochem. Pharmacol. 21, 219 (1972).

GORDON, G. R., A. G. SHAFIZADEH, and J. H. PETERS: Polymorphic acetylation of drugs in rabbits. Xenobiotica 3, 133 (1973).

GRAM, T. E., G. L. LITTERST, and E. G. MIMNAUGH: Enzymic conjugation of foreign chemical compounds by rabbit lung and liver. Drug Metab. Dispos. **2**, 254 (1974).

GREGORY, J. D.: Sulfate conjugation. In B. B. Brodie and E. G. Erdös (eds.): Metabolic factors controlling duration of drug action, vol. 6, p. 53. New York: Pergamon Press (1962).

——, and P. W. ROBBINS: Metabolism of sulfur compounds (sulfate metabolism). Ann. Rev. Biochem. **29**, 347 (1960).

GROVER, P. L.: Conjugations with glutathione. In D. V. Parke and R. L. Smith (eds.): Drug metabolism from microbe to man, p. 105. London: Taylor & Francis Ltd. (1977).

GRÜNOW, W., and C. BOEHME: Uber den Stoffwechsel von 2,4,5-T und 2,4-D bei Ratten und Mäusen. Arch. Toxicol. **32**, 217 (1974).

HARTIALA, K.: Metabolism of hormones, drugs, and other substances by the gut. Physiol. Rev. **53**, 496 (1973).

—— Metabolism of foreign substances in the gastrointestinal tract. In D. H. K. Lee, H. L. Falk, S. D. Murphy, and S. R. Geiger (eds.): Reactions to environmental agents, section 9. Handbook of physiology, p. 375. Bethesda, MD: Amer. Physiol. Soc. (1977).

HATHWAY, D. E.: Foreign compound metabolism in mammals, vols. 1, 2, and 3. London: The Chemical Society, Burlington House (1970, 1972, and 1975).

HEARSE, D. J., and W. W. WEBER: Multiple N-acetyltransferases and drug metabolism, tissue distribution, characterization and significance of mammalian N-acetyltransferase. Biochem. J. **132**, 519 (1973).

HIRAM, P. C., J. R. IDLE, and P. MILLBURN: Some aspects of the biosynthesis and excretion of xenobiotic conjugates in mammals. In D. V. Parke and R. L. Smith (eds.): Drug metabolism—From microbe to man, p. 299. London: Taylor & Francis Ltd. (1977).

HOLLINGSWORTH, R. M.: Biochemistry and significance of transferase reactions in the metabolism of foreign chemicals. In: Reactions to Environmental Agents, Section 9. Handbook of Physiology, p. 455. Bethesda, MD: Amer. Physiol. Soc. (1977).

HUCKER, H. B.: Intermediates in drug metabolism reactions. Drug Metab. Rev. **2**, 33 (1973).

HUNTER, J., and L. F. CHASSEAUD: Clinical aspects of microsomal enzyme induction. In J. W. Bridges and L. F. Chasseaud (eds.): Progress in drug metabolism, vol. 1, p. 129. New York: Wiley (1976).

HUTSON, D. H.: Mechanisms of biotransformation. In D. E. Hathway, Senior Reporter: Foreign compound metabolism in mammals, vol. 1, p. 314. London: The Chemical Society, Burlington House (1970).

—— Mechanisms of biotransformations. In D. E. Hathway, Senior Reporter: Foreign compound metabolism in mammals, vol. 2, p. 328. London: The Chemical Society, Burlington House (1972).

—— Mechanisms of biotransformations. In D. E. Hathway, Senior Reporter: Foreign compound metabolism in mammals, vol. 3, p. 449. London: The Chemical Society, Burlington House (1975).

—— Glutathione conjugates. In D. D. Kaufman, G. G. Still, G. D. Paulson, and S. D. Bandal (eds.): Bound and conjugated pesticide residues, p. 103. Washington, D. C.: American Symposium Series 29, Amer. Chem. Soc. (1976).

IQBAL, Z. M., and R. E. MENZER: Metabolism of O-ethyl-S-S-dipropylphosphorodithioate in rats and liver microsomal systems. Biochem. Pharmacol. **21**, 1569 (1972).

IRVING, C. C.: Conjugates of N-hydroxy compounds. In W. H. Fishman (ed.): Metabolic conjugation and metabolic hydrolysis, vol. 1, p. 53. New York: Academic Press (1970).

—— Metabolic activation of N-hydroxy compounds by conjugation. Xenobiotica **1**, 387 (1971).

—— Interactions of chemical carcinogens with DNA. In H. Busch (ed.): Methods in cancer research, p. 189 (1973).

JAKOBY, W. B.: The glutathione S-transferases A group of multifunctional detoxification proteins. Adv. Enzymol. **46**, 383 (1978).
——, W. H. HABIG, J. H. KEEN, J. N. KETLEY, and M. J. PABST: Glutathione S-transferases: Catalytic aspects. In I. M. Arias and W. B. Jakoby (eds.): Glutathione: Metabolism and function, p. 189. New York: Raven Press (1976 a).
——, J. N. KETLEY, and W. H. HABIG: Rat glutathione S-transferases: Binding and physical properties. In M. Arias and W. B. Jakoby (eds.): Glutathione metabolism and function, p. 213. New York: Raven Press (1976 b).
JAMES, M. O., R. L. SMITH, R. T. WILLIAMS, and M. REIDENBERG: The conjugation of phenyl acetic acid in man, subhuman primates and some primate species. Proc. Roy. Soc. London, Ser. B. **182**, 25 (1972).
JAVITT, N. B.: Biochemical probes for the study of binding and conjugation of glutathione S-transferases. In I. M. Arias and W. B. Jakoby (eds.): Glutathione: Metabolism and function, p. 309. New York: Raven Press (1976).
JERINE, D. M.: Products, specificity and assay of glutathione S-transferase with epoxide substrates. In I. M. Arias and W. B. Jakoby (eds.): Glutathione: Metabolism and function, p. 267. New York: Raven Press (1976).
JERINE, D. M.: and J. W. DALY: Arene oxides: A new aspect of drug metabolism. Science **185**, 573 (1974).
JONES, A. R.: The metabolism of biological alkylating agents. Drug Metab. Rev. **2**, 71 (1973).
JORDON, B. J., and M. J. RANCE: Taurine conjugation of fenclofenac in the dog. J. Pharm. Pharmacol. **26**, 359 (1974).
KADLUBAR, F. F., J. A. MILLER, and E. C. MILLER: Hepatic microsomal N-glucuronidation and nucleic acid binding of N-hydroxy arylamines in relation to urinary bladder carcinogenesis. Cancer Res. **37**, 805 (1977).
KAPLOWITZ, N., and G. CLIFTON: The glutathione S-transferases in rat liver and kidney: Drug induction, hormonal influences and organic anion-binding. In I. M. Arias and W. B. Jakoby (eds.): Glutathione: Metabolism and function, p. 301. New York: Raven Press (1976).
KATZ, R., and A. E. JACOBSON: Chemical structure-activity correlation in the O-methylation of substituted catechols by catechol O-methyltransferase. Mol. Pharmacol. **8**, 594 (1972).
KENNEDY, K. A., K. A. HALMI, and L. J. FISCHER: Urinary excretion of a quaternary ammonium glucuronide of cyproheptadine in humans undergoing chronic drug therapy. Life Sci. **21**, 1813 (1978).
KIESE, M.: The biochemical production of ferrihemoglobin-forming derivative from aromatic amines, and mechanisms of ferrihemoglobin formation. Pharmacol. Rev. **18**, 1091 (1966).
KING, C. M., and C. W. OLIVE: Comparative effects of strain, species, and sex on the acetyltransferase- and sulfotransferase-catalyzed activations of N-hydroxy-N-2-fluorenylacetamide. Cancer Res. **35**, 906 (1975).
KOBAYASHI, K., H. AKITAKE, C. MATSUDA, and S. KUMURA: Studies on the metabolism of chlorophenols in fish. V. Isolation and identification of a conjugated phenol excreted by goldfish, Bull. Japan. Soc. Scient. Fisheries **41**, 1277 (1975).
——, S. KIMURA, and H. AKITAKE: Studies on the metabolism of chlorophenols in fish. VII. Sulfate conjugation of phenol and PCP by fish livers. Bull. Japan. Soc. Scient. Fisheries **42**, 171 (1976).
KURYZNSKE, J. S., and J. T. SMITH: A relationship between the dietary history of a rat and the relative conjugation of glucuronate and sulfate with salicylamide. Fed. Proc. **34**, 882 (1975).
LABOW, R. S., and D. S. LAYNE: A comparison of glucoside formation by liver preparations from the rabbit and the mouse. Biochem. J. **142**, 75 (1974).
LADU, B. N., H. G. MANDEL, and E. L. WAY: Fundamentals of drug metabolism and drug disposition. Baltimore: William & Wilkins (1971).
LADURON, P. M., W. R. GOMMEREN, and J. E. LEYSEN: N-Methylation of biogenic amines. I. Characterization and properties of an N-methyltransferase in rat

brain using 5-methyltetrahydrofolic acid as the methyl donor. Biochem. Pharmacol. 23, 1599 (1974).

LAMOUREUX, G., and K. L. DAVISON: Mercapturic acid formation in the metabolism of propachlor, CDAA, and Fluorodifen in the rat. Pest. Biochem. Physiol. 5, 497 (1975).

LAYNE, D. W.: New metabolic conjugates of steroids. In W. D. Fishman (ed.): Metabolic conjugation and metabolic hydrolysis, vol. 1, p. 22. New York: Academic Press (1970).

LEE, D. H. K., H. L. FALK, S. D. MURPHY, and S. R. GEIGER: Reactions to environmental agents. Handbook of physiology, section 9. Bethesda, MD: Amer. Physiol. Soc. (1977).

LEVY, G.: Dose dependent effects in pharmacokinetics. In D. H. Tedeschi and R. E. Tedeschi (eds.): Importance of fundamental principles in drug evaluation, p. 141. New York: Raven Press (1968).

——— Drug biotransformation interactions in man: Nonnarcotic analgesics. Ann. N.Y. Acad. Sci. 179, 32 (1971).

———, T. TSUCHIYA, and L. P. AMSEL: Limited capacity for salicyl phenolic glucuronide formation and its effect on the kinetics of salicylate elimination in man. Clin. Pharmacol. Therap. 13, 258 (1972).

———, B. YAGEN, and R. MECHOVLAM: Identification of a C-glucuronide of Δ^6-tetra-hydrocannabinol as a mouse liver conjugate, in vivo. Science 200, 1931 (1978).

LISTOWSKY, I., K. KAMISAKA, K. ISHITANI, and I. M. ARIAS: Structures and properties of ligandin. In I. M. Arias and W. B. Jakoby (eds.): Glutathione: Metabolism and function, p. 233. New York: Raven Press (1976).

LOTLIKAR, P. D.: Enzymatic N-O-methylation of hydroxyamic acids. Biochim. Biophys. Acta. 170, 468 (1968).

———, and M. B. WASSERMAN: Reactive phosphate ester of the carcinogen 2-(N-hydroxy)acetamido-fluorene. Biochem. J. 120, 661 (1970).

McBAIN, J. B., and J. J. MENN: S-Methylation, oxidation, hydroxylation and conjugation of thiophenol in the rat. Biochem. Pharmacol. 18, 2282 (1969).

MEISTER, A.: Glutathione synthesis. In P. H. Boyer (ed.): The enzymes, vol. 10, p. 671. New York: Academic Press (1974).

——— Biochemistry of glutathione. In D. M. Greenberg (ed.): Metabolism of sulfur compounds, 3rd ed., p. 101. New York: Academic Press (1975).

MIETTINEN, T. A., and E. LESKINEN: Glucuronic acid pathway. In W. H. Fishman (ed.): Metabolic conjugation and metabolic hydrolysis, vol. 1, p. 157. New York: Academic Press (1970).

MILLBURN, P.: Factors affecting glucuronidation in vivo. Biochem. Soc. Trans. 2, 1182 (1974).

MILLER, E. C., and J. A. MILLER: Biochemical mechanisms of chemical carcinogenesis. In H. Bush (ed.): The molecular biology of cancer, p. 377. New York: Academic Press (1974).

MILLER, J. A., and E. C. MILLER: Metabolic activation of carcinogenic aromatic amines and amides via N-hydroxylation and N-hydroxyesterification and its relationship to ultimate carcinogens as electrophilic reactants. In E. D. Bergmann and B. Pulman (eds.): Physiochemical mechanisms of carcinogenesis, vol. 1, p. 237. New York: Academic Press (1969).

——— ——— The concept of reactive electrophilic metabolites in chemical carcinogenesis: Recent results with aromatic amines, safrole and aflatoxin B_1. In D. J. Jollow, J. J. Kossis, R. Snyder, and H. Vainio (eds.): Biological reactive intermediates-Formation, toxicity and inactivation, p. 6. New York: Plenum Press (1977).

MILLER, J. J., G. M. POWELL, A. H. OLAVESEN, and C. G. CURTIS: The fate of 2,6-dimethoxy [U-^{14}C] phenol in the rat. Xenobiotica 4, 285 (1974).

MINCK, K., R. R. SCHUPP, H. P. A. ILLING, G. F. KAHL, and K. J. NETTER: Interrelationship between dimethylation of p-nitroanisole and conjugation of p-nitrophenol in rat liver. Naunyn-Schmiedebergs Arch. Pharmacol. 279, 347 (1973).

70   G. D. PAULSON

MITCHELL, J. R., J. A. HINSON, and S. D. NELSON: Glutathione and drug-induced
    tissue lesions. In I. M. Arias and W. B. Jakoby (eds.): Glutathione: Metabolism
    and function, p. 357. New York: Raven Press (1976).
MOORE, D. H.: Species, sex and strain differences in metabolism. In D. E. Hathway
    (ed.): Foreign compound metabolism in mammals, vol. 2, p. 398. London Chemi-
    cal Society (1972).
MUDD, S. H.: Biochemical mechanisms in methyl group transfer. In W. H. Fishman
    (ed.): Metabolic conjugation and metabolic hydrolysis, vol. 3, p. 297. New York:
    Academic Press (1973).
MULDER, G. J., and A. H. E. PILON: UDP glucuronyltransferase and phenolsulfotrans-
    ferase from rat liver in vivo and in vitro. III. The effect of phenolphthalein and
    its sulfate and glucuronide conjugate on conjugation and biliary excretion of
    harmol. Biochem. Pharmacol. 24, 517 (1975).
OEHME, F. W.: Species differences: The basis for and importance of comparative
    toxicology. Clin. Toxicol. 3, 5 (1970).
PARKE, D. V.: The biochemistry of foreign compounds. Internat. Ser. Monographs Pure
    & Applied Biol. Biochem. Div. New York: Pergamon Press (1968).
PARKE, D. F., and R. L. SMITH (eds.): Drug metabolism—From microbe to man.
    London: Taylor and Francis Ltd. (1977).
PAULSON, G. D.: Sulfate ester conjugates—their synthesis, purification, hydrolysis, and
    chemical spectral properties. In D. D. Kaufman, G. G. Still, G. D. Paulson, and
    S. D. Bandal (eds.): Bound and conjugated pesticide residues, p. 86. Washing-
    ton, D. C.: Amer. Chem. Soc. Symposium Series (1976).
——, and M. V. ZEHR: Metabolism of p-chlorophenyl N-methylcarbamate in the
    chicken. J. Agr. Food Chem. 19, 471 (1971).
PECK, H. D.: Sulfation linked to ATP cleavage. In P. D. Boyer (ed.): The enzymes,
    vol. 10, p. 651. New York: Academic Press (1974).
PORTER, C. C., B. H. ARISON, V. F. GRUBER, D. C. FLUS, and W. J. A. VANDENHEUVAL:
    Human metabolism of cyproheptadine. Drug Metab. Dispos. 3, 189 (1975).
RADOMSKI, J. L., W. L. HEARN, T. RADOMSKI, H. MORENO, and W. E. SCOTT: Isola-
    tion of the glucuronic acid conjugate of N-hydroxy-4-aminobiphenyl from dog
    urine and its mutagenic activity. Cancer Res. 37, 1757 (1977).
REMY, C. N.: Metabolism of thiopyrimidines and thiopurines: S-Methylation with S-
    adenonylethionine transmethylase and catabolism in mammalian tissues. J. Biol.
    Chem. 238, 1078 (1963).
RICHTER, W. J., K. O. ALT, W. DIETERLY, J. W. FAIGLE, H. P. KRIEMLER, H. MORY
    and T. WINKLER: C-Glucuronides, a novel type of drug metabolites. Helv. Chim.
    Acta 58, 2512 (1975).
ROBBINS, P. W.: Sulfate transfer. In P. D. Boyer, H. Lardy, and K. Myrbäck (eds.):
    The enzymes, vol. 6, 2nd ed., p. 363. New York: Academic Press (1962).
ROY, A. B.: The synthesis and hydrolysis of sulfate esters. Adv. Enzymol. 22, 205
    (1960).
—— Sulphate conjugation enzymes. Handbüch der Experimentellen Pharmakologie
    28, 536 (1971).
SARRIF, A. M., and C. HEIDELBERGER: On the interaction of chemical carcinogens with
    soluble proteins of target tissues and in cell culture. In I. M. Arias and W. B.
    Jakoby (eds.): Glutathione: Metabolism and function, p. 317. New York: Raven
    Press (1976).
SATO, T., T. SUZUKI, and T. FUKUYAMA: Studies on conjugation of [^{35}S] sulfate with
    phenolic compounds. IV. Metabolism of o-cresol, m-cresol, salicyladehyde,
    salicylic acid, toluene, benzoic acid and related substances in rat's liver. J. Bio-
    chem. 43, 421 (1956).
SIEBER, S. M., and R. H. ADAMSON: The metabolism of xenobiotics by fish. In D. W.
    Parke and R. L. Smith (eds.): Drug metabolism—From microbe to man, p. 233.
    London: Taylor & Francis Ltd. (1977).
SKLAN, N., and E. A. BARNSLEY: The metabolism of S-methyl-L-cysteine. Biochem. J.
    107, 217 (1968).

SLOTKIN, T., and V. DISTEFANO: Urinary metabolites of harmine in the rat and their inhibition of monoanine oxidase. Biochem. Pharmacol. **19**, 125 (1970).

SMITH, J. N.: Detoxication mechanisms in insects. Biol. Rev. Cambridge Philos. Soc. **30**, 455 (1955).

—— The comparative metabolism of xenobiotics. Adv. Comp. Physiol. Biochem. **3**, 173 (1968).

—— Comparative detoxication of invertebrates. In D. V. Parke and R. L. Smith (eds.): Drug metabolism—From microbe to man, p. 219. London: Taylor & Francis Ltd. (1977).

SMITH, R. G., G. C. DAVES, JR., R. K. LYNN, and N. GERBER: Hydantoin ring glucuronidation: Characterization of a new metabolite of 5,5-diphenylhydantoin in man and the rat. Biomed. Mass Spec. **4**, 275 (1977).

SMITH, R. L., and J. CALDWELL: Drug metabolism in non-human primates. In D. V. Parke and R. L. Smith (eds.): Drug metabolism—From microbe to man, p. 331. London: Taylor & Francis Ltd. (1977).

——, and R. T. WILLIAMS: Implication of the conjugation of drugs and other exogenous compounds. In G. J. Dutton (ed.): Glucuronic acid, p. 457. New York: Academic Press (1966).

—— History of the discovery of the conjugation mechanisms. In W. H. Fishman (ed.): Metabolic conjugation and metabolic hydrolysis, vol. 1, p. 1. New York: Academic Press (1970).

SULLIVAN, H. R., S. L. DUE, and R. E. MCMAHON: Metabolism of α-1-methadol; N-acetylation, a new metabolic pathway. Res. Comm. Chem. Pathol. Pharmacol. **6**, 1072 (1973).

TATE, S. S., G. A. THOMPSON, and A. MEISTER: Recent studies on γ-glutamyl transpeptidase. In M. Arias and W. B. Jakoby (eds.): Glutathione: Metabolism and function, p. 45. New York: Raven Press (1976).

USDIN, E., and S. H. SNYDER: Frontiers in catacholamine research. New York: Pergamon Press (1973).

VAINIO, H., and A. AITTIO: Enhancement of microsomal drug hydroxylation and glucuronidation in rat liver by phenobarbital and 3-methylcholanthrene in combination. Acta Pharmacol. Toxicol. **34**, 130 (1974).

VAISMAN, S. L., and K. S. LEE, and L. M. GARTNER: Xylose, glucose, and glucuronic acid conjugation in the newborn rat. Pediatr. Res. **10**, 967 (1976).

VESSEY, D. A., and D. ZAKIM: The identification of a unique p-nitrophenol conjugating enzyme in guinea pig liver microsomes. Biochim. Biophys. Acta **315**, 43 (1973).

VESTERMARK, A., and H. BOSTRÖM: On the sulphurylation of mono- di- and trihydric phenols. Experientia **16**, 408 (1960).

WAKABAYASHI, M.: β-Glucuronidases in metabolic hydrolysis. In W. H. Fishman (ed.): Metabolic conjugation and metabolic hydrolysis, vol. 2, p. 520. New York: Academic Press (1970).

WEBER, W. W.: Acetylating, deacetylating and amino acid conjugating enzymes. In B. B. Brodie and J. R. Gillette (eds.): Handbook of experimental pharmacology. Concepts in biochemical pharmacology II, vol. 28, part 2, p. 564. New York: Springer-Verlag (1971).

—— Acetylation of drugs. In W. H. Fishman (ed.): Metabolic conjugation and metabolic hydrolysis, vol. 3, p. 249. New York: Academic Press (1973).

WEISSBURGER, J. H., R. S. YAMAMOTO, G. M. WILLIAMS, P. H. GRANTHAM, T. MATSUSHIMA, and E. K. WEISBURGER: On the sulfate ester of N-hydroxy-N-2-fluorenylacetamide as a key ultimate hepatocarcinogen in the rat. Cancer Res. **32**, 491 (1972).

WENGLE, B.: Studies on ester sulphates; on sulphate conjugation in foetal human tissue extracts. Acta Soc. Med. Upsalien **69**, 105 (1963).

WILLIAMS, F. M., R. H. BRIANT, C. T. DOLLERY, and D. S. DAVIES: The influence of the route of administration on urinary metabolites of isoetharine. Xenobiotica **4**, 345 (1974).

72 G. D. PAULSON

WILLIAMS, R. T.: Detoxication mechanisms, 1st and 2nd eds. London: Chapman & Hall (1947 and 1959).
—— The biogenesis of conjugation and detoxication products. In P. Bernfield (ed.): Biogenesis of natural compounds, 2nd ed., p. 589. New York: Pergamon Press (1967).
—— Introduction: Pathways of drug metabolism. In B. B. Brodie and J. R. Gillette (eds.): Concepts in biochemical pharmacology, part II, p. 233. New York: Springer-Verlag (1971).
—— Inter-species variation in the metabolism of xenobiotics. Biochem. Soc. Trans. 2, 359 (1974).
WIT, J. G.: Drug metabolism in avian species. In D. V. Parke, and R. L. Smith (eds.): Drug metabolism—From microbe to man, p. 247. London: Taylor & Francis Ltd. (1977).
WOOD, J. L.: Biochemistry of mercapturic acid formation. In W. H. Fishman (ed.): Metabolic conjugation and metabolic hydrolysis, vol. 2, p. 261. New York: Academic Press (1970).
YAGER, B., S. LEVY, and R. MECHOVLAM: Synthesis and enzymatic formation of C-glucuronide of Δ^6-tetrahydrocannabinol. J. Amer. Chem. Soc. 99, 6444 (1977).
YOUNG, L.: The metabolism of foreign compounds—History and development. In D. V. Parke and R. L. Smith (eds.): Drug metabolism—From microbe to man, p. 1. London: Taylor & Francis Ltd. (1977).

Manuscript received December 27, 1979, accepted December 31, 1979.

Microbial agents as insecticides

By

JOHN W. CHERWONOGRODZKY*

Contents

* Department of Microbiology and Parasitology, University of Toronto, Toronto, Ontario M5S 1A1, Canada.

© 1980 by Springer-Verlag New York Inc.
Residue Reviews, Volume 76

I. Introduction

In the competition between man and insects for limited food resources, at times it appears that the latter group has the upper-hand. Not only do they have far shorter generation times and staggering numbers of off-spring, but the tendency for agricultural businesses to specialize in only a few crops has worked to their advantage. Rather than succumbing to starvation or predation in their search for suitable food sources, they are now given several acres at a time to populate (ADAMS *et al.* 1971). Over the years, chemical insecticides have proven useful in lowering their numbers, but even these are now being found to have several limitations:

1. Insects are acquiring resistance to chemical pesticides not only through increased individual tolerances (CROW 1957) but also by the selection of resistant progeny (MARTIGNONI and SCHMID 1961, WHITTEN 1978, PRIESTER and GEORGHIOU 1979).

2. General pesticides may eliminate economically beneficial insects as well as the natural predators of those that are destructive (SMITH and VAN DEN BOSCH 1967).

3. As in the example of DDT, several have come into disfavor due to their persistence in the environment and accumulation in the food-chain by biological concentration (DUSTMAN and STICKEL 1969).

4. They may be inactivated by weather, soil conditions, or biodegradation (LICHTENSTEIN 1966).

5. The cost of production or application of the insecticide may be inhibitory (HOWSE 1973).

6. Several are toxic to human beings. In the case of seed grain contaminated with hexachlorobenzene in Turkey, more than 3,000 became ill and numerous deaths were reported after eating this formulation (*Secretary's Comission on Pesticides* 1969).

It is because of the above limitations that other alternatives for controlling insect populations are being investigated. Of these, microbial agents have proven successful and indeed their industrial production, patents, distribution, and toxicological studies resemble those for chemical insecticides (BURGES and HUSSEY 1971 a). When they are applied to a field, however, the questions arise—will the delicate ecological balances be overturned, and will the disease spread to other lifeforms?

With regard to the use of microbes upsetting ecological balances, to a large extent such fears are unfounded. It should be remembered that the use of such agents is merely to correct an already existing imbalance, to lower the numbers of a pest back to a tolerable number (HUFFAKER *et al.* 1976). When faced with the loss of millions of acres of forest (STAIRS 1971) or the total elimination of a country's cash crop within a few years (EASWARAMOORTHY and JAYARAJ 1978), these microorganisms have in fact protected the environment from other desperate measures such as massive spraying programs with toxic chemicals. It is unfortunate that, as yet, microbial agents are still being overshadowed by the use of such chemical pesticides (BURGES and HUSSEY 1971 b).

As for the microorganisms infecting lifeforms other than their intended targets, in several instances their specificities offer distinct advantages over chemical insecticides. *Bacillus popillae*, for example, can infect only certain closely related beetles of the family Scarabaeidae (DUTKY 1963) while, for several of the insect viruses, they usually do not affect more than one insect in a mixed population (STAIRS 1971).

Despite these merits, it is unwise to make assumptions. Unlike chemical pesticides, microbial agents are living systems that can replicate and hence persist for several generations in the environment. In one context, this is an advantage. Fields can be protected against a certain pest for some time after they have been applied. In another context, however, not all microbial agents are host specific. The Nodamura virus, for example, not only fatally infects honeybees and the greater wax moth, but it also grows in mammalian cell cultures and kills suckling mice (BAILEY *et al.* 1975). It is because of this diversity that a general survey is now presented. As the topic is far too complex and detailed for the limits of this review, only a few representatives of each group of microbial agents will be given. In this manner, it is hoped to inform rather than to overwhelm the reader.

II. Fungi

a) Background

The earliest recorded history of insect diseases begins with Aristotle (384 to 322 BC). In his *Historia Animalium*, he described his observations on the diseases of honeybees. Later, in the sixteenth century AD, several reports were made on the diseases of silkworms although it was not until 1834 that Agostino Bassi showed that these could be caused by microorganisms. In his studies he found that the muscardine disease of silkworms was caused by the fungus *Beauveria bassiana* (BURGES and HUSSEY 1971 a).

Widely distributed, fungi are a constant factor in the natural control of insects. Within Canada, for example, the *Entomophthora* species has had dramatic effects. In the early 1960s, the grasshopper invasion of western Canada collapsed due to the infections by *Entomophthora grylli*. Sampling studies showed that 95% of the grasshoppers had succumbed to this disease (see Figs. 1 a and b). On a similar trend is the decline of the eastern hemlock looper in Newfoundland due to the spread of *Entomophthora egressa* and *Entomophthora sphaerosperma* (TYRRELL and MacLEOD 1973). On an international level is the concern over the spread of disease by insect vectors. As these carriers are becoming increasingly more resistant to chemical pesticides, the fungi, *Coelomomyces*, which are obligate parasites of mosquitoes, sandflies, blackflies, and midges are

(a)   (b)

Fig. 1a. Grasshoppers, *Melanoplus bivittatus* Say, destroyed by the fungus *Ento-mophthora grylli* Fresenius. Several typical symptoms of the disease are evident— congregation in a vertical position near the top of the plant stem, clasping, and partial disintegration of some insects (× 1.5).

Fig. 1b. Resting spores of the fungus taken from the body cavity of the grasshoppers (× 700).

(Photographs by Dr. D. MacLeod, Great Lake Research Centre and Insect Pathology Research Institute, Sault Ste. Marie, Ontario.)

currently being researched as alternatives (Laird 1971). Fungi, therefore, are effective agents for insect control.

b) Mode of infection

As all fungi must obtain their organic requirements from external sources, the majority are saprophytes, feeding upon the products or remains of plants and animals. Several, however, parasitize other living organisms for their own needs (Alexopoulos 1962). In many instances the insects are the ones that are the susceptible hosts.

Although infection may occur through the respiratory or alimentary tract (Veen 1966, Yendol and Paschke 1965), this seldom happens. The usual method is for the conidium or fungal spore to germinate on the cuticle. The cuticle may have such antifungal agents as medium-chain saturated fatty acids (KOIDSUMI 1957), but any abrasion or damage to this layer will lower this resistance. Other compounds such as lipids in the cuticle or the underlaying fat layer tend to enhance germination (ROBERTS and YENDOL 1971). Although warm, moist environments assist the initiation of infection, there are exceptions. The cockroach lives in a habitat that should be ideal for fungal disease, yet it is relatively disease-free (ROBERTS and YENDOL 1971).

After germination the mycelium penetrates through the tissues by the release of enzymes and mechanical force. In most cases, yeast-like fragments of mycelia are produced which then float free to multiply in the haemocoel (PRASERTPHON and TANADA 1968). Death of the host then occurs either by the release of fungal toxins or by the destruction of the tissues. Invasion continues until the insect is filled with mycelia. Conidiophores then erupt through the cuticle and produce spores to continue the cycle (BELL 1974).

c) Field trials

In 1890 the state of Kansas (U.S.A.) provided preparations of *Beauveria bassiana* to combat the chinch bug, *Blissus leucopterus*, which attacked the cereal grain crops. The first two years showed some reduction in the numbers of the pest, but in subsequent years applications of the fungus proved to be of little effect. The failure of the program led to two conclusions, the first being that the mere presence of a microbial control agent is not enough to eliminate exposed insects, and the second being that environmental conditions such as the weather are key factors in its effectiveness (ROBERTS and YENDOL 1971).

In a more recent study, *Cephalosporium lecanii*, the white halo fungus, was tested against the coffee green bug, *Coccus viridis*. In southern India where coffee is an important cash crop, the insect infests the underside of the leaf, causing defoliation and fruit drop. Isolated from infected green bugs, the fungus was then grown on agar slants or moist sterile sorghum grain, filtered, and the spores sprayed on the bushes at two-week intervals. On the second application, mortality was estimated at 73%. If 0.05% Tween 20 was added, the mortality increased to 98%. If the environment was dry, its effects were minimized by adding glycerol as a humectant to the spore preparation (EASWARAMOORTHY and JAYARAJ 1978).

As fungal insect pathogens are widely distributed in nature and often have variable effectiveness, generally they have not been pursued as commercial products. The one exception has been the factory at Krasmodar in the U.S.S.R. which mass-produces *Beauveria bassiana* under the name Beauverin (VAN DER GEEST and VAN DER LAAN 1971).

d) Toxicity and pathogenicity

The effects of the insect fungi on their hosts have already been described; however, the effects on other animal species will determine the degree of safety or whether these agents are sensible alternatives to chemical insecticides.

1. Aflatoxins.—Members of the *Aspergillus* species, notably *A. flavus* and *A. parasiticus*, produces several aflatoxins. In feeding experiments against the boll weevil, 0.6 ppm killed 36% of the animals within a week, 0.06 ppm caused sterility, while 0.006 ppm was ineffective (MOORE *et al.* 1978). Due to the controversy of these aflatoxins, however, it is difficult to determine the practicality of these as insecticides. Although rats, trout, ducks, and sheep are sensitive to these compounds, developing carcinomas and hepatomas, human beings and mice are highly resistant to the effects (WOGAN 1968). Possibly this difference stems from variations in metabolism, the former group activating aflatoxin B_1 to a carcinogen (PATTERSON 1973). At present there is only tentative evidence that such agents produce liver cancer in man (KRAYBILL and SHAPIRO 1969), while the presence of other mycotoxins may also be significant (GOLDBLATT 1969).

2. Infections.—Although fungi are responsible for several notorious diseases in man and livestock (ALEXOPOULOS 1962), the insect fungi have been evident only in a few isolated cases. Within horses, *Entomophthora coronata* has been found to cause ulcerated nodules in greatly distorted nostrils and upper lips (BRIDGES *et al.* 1962). In the tropics, this same agent has caused subcutaneous nodules and severe facial edema in man. These cases are believed to be exceptions with the fungus taking advantage of a host with temporarily lowered resistance (BRAS *et al.* 1965, RESTREPO *et al.* 1967). When *Beauveria bassiana* Vuillemin was injected subcutaneously into white mice, most succumbed to infections in the lung, brain, and kidney. Inhalation of the spores proved harmless (MACLEOD 1954). Indeed, DIEUZEIDE in 1925 recommended taking *Beauveria* spores with a little water and tea to remedy infantile meningitis and epilepsy.

As for the effect on crops, when spore dust of *Beauveria bassiana* Vuillemin was applied to grasses, weeds, and oranges, there was little pathogenicity. Direct inoculation into tomatoes, however, resulted in spoilage (MACLEOD 1954).

In laboratory tests *Beauveria bassiana* could infect bees. Despite its widespread occurrence in the insect order, however, it has never been found attacking bees in nature (CANTWELL *et al.* 1966). Rather than upsetting the ecological balance, the *Entomophthora* species appears to be an integral part of one. It infects most, if not all, of the insects in nature and can be transmitted to widely different host (YENDOL and PASCHKE 1966).

3. Toxicity.—In 1968, SCHAERFFENBERG (as reported by HEIMPEL 1971) conducted investigations on *Beauveria bassiana* and *Metarrhizium anisopliae*. The results were:

1. Subcutaneous and intraveneous injections of 0.5 ml of a 2% spore suspension were given to 2 groups of 10 rats each. No morbid changes or invasions of the blood and organs were found after 2 mon.

2. In inhalation experiments, 2 groups of 10 rats were exposed to 4

mg of spore dust/m^3 in a closed container. None showed pathogenic, toxic, or allergic effects.

3. Three groups of 10 rats each were fed different formulations. The group that received 6 g of fungus mixed 1:1 with sugar and rolled oats, and then was fed bread, fruit, and vegetables, showed no effects. The second group that was fed a constant diet of fungus, bran, rolled oats, and sugar showed a loss of appetite and a weight loss of 30%. Those fed only fungus and bran died in 3 cases, while the rest showed a weight loss of 40%, diarrhea, and stopped feeding.

The above data supported MacLeod's opinion (1954) that animals under stress were susceptible to the action of insect fungi.

For human beings Müller-Kögler (as reported by Heimpel 1971) noted that scientists working with the spores of *Beauveria bassiana* developed moderate to severe allergic reactions including neck or head pains, giddiness, and fatigue. The symptoms usually disappeared in a short time and could be avoided by wearing masks and protective clothing.

III. Protozoa

a) Background

Unicellular eukaryotes, the protozoa, for the most part are free-living organisms found in soil and water. Although they usually feed on smaller prey such as bacteria or algae, several are parasitic to higher forms of life. Those that are pathogenic to insects are diverse in both their virulence and their morphology but can be grouped into 4 general subdivisions (Stanier 1970, McLaughlin 1971):

1. **Mastigophora.**—These are flagellates that are motile with one or more flagellae; their cell division is always longitudinal. Pathogenesis is either mild or undetected, and if death of the host does occur, it is usually by other parasites. Of very little specificity, they infect a wide range of insects and can sometimes interchange among their hosts.

2. **Rhizopoda.**—Amoeboid, they are motile by means of pseudopodia although some species can also form flagellae. Reproduction is by binary fission. *Malpighamoeba mellificae* is a honeybee amoeba and *M. locustae* is a serious pathogen of grasshoppers.

3. **Sporozoa.**—The largest group of entomophilic protozoa, they usually parasitize a wide range of insects. These tend to devitalize or reduce the number of offspring rather than kill their hosts. Cell division is by multiple fission and they are either immobile or show gliding movement.

4. **Ciliates.**—Movement is by coordinated beating of numerous cilia. The cell has 2 nuclei differing in structure and function. Cell division is transverse. Few ciliate infections are found in nature, although *Tetrahymenidae stegomyiae* has been shown to prevent maturation of mosquito larvae after infection under laboratory conditions (Kellen et al. 1966).

b) Mode of infection

The *Tetrahymenidae* ciliates are distinct from the other species in that they have an invasive ability. After penetrating the cuticle of mosquito larvae, they then multiply rapidly for 2 days until their numbers level off. The host larvae can live for some time though they seldom mature to the pupa or adult stage (KELLEN *et al.* 1966).

For most other protozoa, the primary route is the alimentary tract after ingestion. In some cases transovum (eggs contaminated on the exterior) or transovarial (congenital) transmission, as well as by other parasites, may occur. Once thought to be highly host specific, most are now known to be able to cross-infect several orders of insects. Seldom virulent, most simply diminish the host's chances of survival by lowering vitality, progeny number, life-span, and responses of stimuli (McLAUGHLIN 1971).

c) Field trials

Due to the attraction of other pathogen-host systems, there have been few attempts reported in the literature to use protozoa as microbial insecticides. In 1956, WEISER and VEBER sprayed fruit trees with a spore suspension of the microsporidum, *Thelohania hyphantriae*, in the hope of controlling infestations by the fall webworm, *Hyphantria cunea*. Although the larvae did succumb to the protozoan, when they died or were selectively preyed upon because of their diminished responses, the microsporidum was also removed from the population.

Greater success has been noted when feeding-baits contaminated with protozoa were used. The boll weevil, *Anthonomus grandis*, feeds within the cotton plant fruit and so even complete spray coverage can at times be ineffective. If bait with *Glugea gasti* is left in the field for those insects preparing to overwinter, infection can be as high as 80 to 98% with the average winter mortality in the treated area being 96% (McLAUGHLIN *et al.* 1969).

In a similar manner, spores of another microsporidum, *Nosema locustae*, were tested against grasshoppers in a comparison of ultra-low volume (ULV) applications with feeding-bait of 10^9 spores/1.5 lb of bran/A. Although both were effective, causing roughly a 50% reduction in grasshopper numbers, the ULV application had to be applied more frequently and at higher spore concentrations/A than did the wheat bran formulation (HENRY *et al.* 1978).

d) Toxicity and pathogenicity

As large scale investigations of protozoa as control agents are few, at present there are no toxicological studies for such discussion. Under pathogenicity, however, it has been shown that the microsporidum, *Thelohania pristiphorae*, from the larch sawfly can infect the completely unrelated tent caterpillar after ingestion (SMIRNOFF 1968). As all beneficial insects, except for the silkworm, dwell in the open, distribution of pathogens of wide host range in the field may infect the former. Honeybees are natural victims and their susceptibility should be tested before any field trial (BAILEY 1971).

Within vertebrates, MATSUBAYASHI et al. (1959) reported the isolation of Encephalitozoan-like bodies from a 9 year-old boy who suffered recurrent fever, headaches, and repeated loss of consciousness. WEISER (1964), however, believed these results were simply due to an experimental error, isolating this species from a latent infection in the mouse that was the initial test animal. PETRI (1966) reported that *Nosema cuniculi* was lethal when injected into mice and produced ascites tumours in the peritoneal cavity. ARISON and CASSARO (1966), in contrast, reported low pathogenicity of a microsporidum in mice and that it retarded the growth of transplantable tumours, prolonged survivals, and even offered some resistance to later transplants. The pathogenicity of insect protozoa in mammals, therefore, is under considerable controversy.

Within fresh- or salt-water lifeforms, infections by protozoa are a natural occurrence. Northern shrimp usually have a 5% loss in egg vitality that was previously assumed to be a deficiency in fertilization. The actual cause has been found to be through infestations by the parasitic dinoflagellate, *Syndinium* (STICKNEY 1978). Smelts (HALEY 1954), haddock (KABATA 1959), and flounders (STUNKARD and LUX 1965) are also susceptible to microsporida infections of the gills and fin muscles. The use of such agents for the control of insect populations, therefore, should also be tested for their effect on other waterlife.

IV. Rickettsiae

Rickettsiae resemble viruses in that they are small, obligate pathogens of animal cells, yet, in several respects, they are also similar to bacteria. Having both DNA and RNA, muramic cell walls, cellular structures, and active metabolic pathways, they are also sensitive to antibiotics (DAVIS et al. 1970). Although most of these organisms are harmless to insects, a few of the *Rickettsiella* species are pathogenic. In the white grubs of lamellicorn beetles, for example, *R. melolonthae* causes "Lorsch disease." After ingesting the microbe by feeding on infected cadavers (NIKLAS 1960), the grubs propagate the pathogen in the cytoplasm of their cells. The first symptoms appear 2 to 3 mon after infection, and about 85% of the grubs die after 4 to 6 mon at 20°C (KRIEG 1963). Changes in behavior or reduction in mobility may be due to replication of the organism in the nervous system of the insect (NIKLAS 1957). The characteristic opalescent discoloration of the diseased larvae is due to the masses of rickettsiae liberated into the hemolymph (KRIEG 1971). Unfortunately, *R. melolonthae* is also pathogenic for mammals such as mice and rabbits (KRIEG 1955, GIROUD et al. 1958, KRIEG 1962), having a high tropism to the lungs of these animals.

Concerning pathogenicity and toxicity of other rickettsiae, although they are frequently pathogenic to vertebrates such as man, they are seldom harmful to the vector. *Rochalimaea quintana* which causes trench fever in man, for example, is not pathogenic to the louse despite its growth in the latter's gut. *Rickettsia typhi*, which causes typhus fever in man, grows intracellularly within the louse gut epithelial cells without pathological effects (Stuart-Harris 1967). It is because of this nature that these agents will probably never be used for control of insects.

V. Viruses

a) Background

When the European spruce sawfly was accidentally introduced into North America during the early part of the twentieth century, within a few years it threatened to destroy the spruce forests of the continent. As a counter-measure, entomologists imported natural parasites of this insect from Europe. A fortunate outcome of this move was that some of these parasites harboured NPV (nuclear polyhedrosis virus) which then swept through the sawfly population to cause a spectacular "insect crash" (Stairs 1971). This success initiated insect viral research in Canada which now undertakes investigations of the 5 main categories (Cunningham 1973):

1. **Nuclear polyhedrosis viruses.**—The most studied group of the insect viruses, Figure 2 shows some of its common features. As can be seen, several rod-shaped virus particles which have DNA as the nucleic acid are found embedded within each inclusion body. The inclusion body is large enough to be seen with a light microscope and is composed primarily of proteins serologically unrelated to the host (Krywienczyk and Bergold 1961). The main tissues infected are the midgut, fat, and epidermal cells.

2. **Granulosis viruses.**—Similar to NPV, they differ in that only one virus particle is found in each inclusion body which is capsule shaped.

3. **Cytoplasmic polyhedrosis viruses.**—These spherical virus particles have RNA as the nucleic acid and infect the cytoplasm of gut cells only.

4. **Entomopoxviruses.**—These are large oval viruses embedded in oval inclusion bodies; their nucleic acid is DNA. These infect the cytoplasm of gut, fat, and epidermal cells.

5. **Nonoccluded viruses.**—A miscellaneous group of varying size, shape, and nucleic acid composition, these are all insect viruses which are not embedded in inclusion bodies.

b) Mode of infection

Despite the diverse forms of insect viruses, infections are usually the same. The insect becomes exposed by the ingestion of contaminated

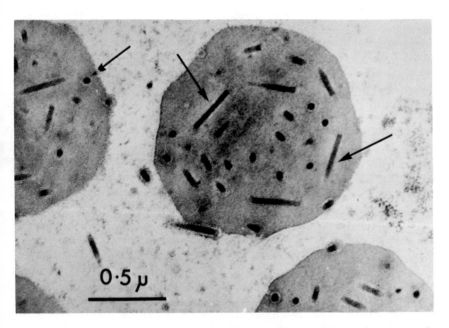

Fig. 2. An electron micrograph of a sectioned polyhedron. Arrows refer to a single
virus rod embedded in the protein matrix (× 62,000).

(Photograph by Dr. J. C. Cunningham, Great Lakes Research Centre and Insect
Pathology Research Institute, Sault Ste. Marie, Ontario.)

soil or foliage, or by transovarial (the virus is within the eggs laid by
chronically infected adults) and transovum transmission (the virus is
on the outside of the egg after fecal contamination). Once inside the
host, the virus then replicates within the tissues, causing cell degen-
eration and disruption. Although death is slow when compared to
chemical pesticides, usually being 20 days or longer, the more virulent
strains may kill a larva within 5 days. Generally, as the insect larva
matures, its susceptibility diminishes, and so preparations of the virus
to contaminate foliage and soils are done early in the season (CUN-
NINGHAM 1973).

c) Field trials

As viruses require living hosts for their replication, synthetic media
cannot be used. Although some success has been reported for cultivation
in cell cultures (VAIL et al. 1973, LYNN and HINK 1978), this is not the
usual procedure. Frequently either laboratory infected insects are released, or the virus is purified from reared larvae. In the latter method,

a week after ingesting the pathogen, the dead or dying hosts are collected and allowed to putrify in water at room temperature. The released
inclusion bodies containing the viruses are resistant to degradation and
settle as a white sediment. The supernatant is poured off and the sediment can be further purified by differential centrifugation (STAIRS
1971). In the winter of 1971 to 1972, for example, the Insect Pathology
Research Institute at Sault Ste. Marie, Ontario, undertook a large-scale
rearing program. Over two million budworm larvae were grown on
artificial diets in plastic cream cups, infected with virus, harvested, and
then processed as mentioned previously (CUNNINGHAM 1973). In a
similar manner, several companies within the United States have produced viral suspensions as commercial products (VAN DER GEEST and
VAN DER LAAN 1971). The yields are roughly 10^{10} polyhedra/larva
(IGNOFFO 1965) while usually 10^9 to 10^{11} polyhedra/A are applied to a
field (STAIRS 1971).

For the method of application, the viral preparations resemble chemical insecticides in that they are either dusted and sprayed by hand over
a few trees, or are dispersed by aircraft and truck-mounted mist-blowers
over large tracts of land. In this manner, several successes have been
noted in controlling sawflies, butterflies, and moths (STAIRS 1971). The
difference from general pesticides is that years later they may still be
effective, propagated by a host target infecting other generations of its
kind. Also, sprayed in one area, the insect may travel to another area,
setting new centers of disease (BIRD and BURK 1961).

The nonoccluded viruses, unlike the above encapsulated strains, cannot be prepared in a similar manner. For the field trial in Florida (U.S.A.)
against the citrus red mite, Panonychus citri, the mites were artificially
infected and then distributed among healthy populations. The disease
then spread to the others, causing paralysis, diarrhea, and death (SMITH
et al. 1959).

Despite these successes, there have been limitations. The target insect must ingest either contaminated foliage or the remains of a previous
host to be affected. In the interim, the virus is exposed to ultraviolet
light, becoming inactivated within 3 to 6 hr if in direct sunlight (DAVID
et al. 1968). Also, not all field trials have proved successful even with
infective agents. While several strains of NPV isolated from sawflies
are effective against their respective hosts, that of the Jack pine sawfly
are not. On trees drenched with suspensions of the viral concentrate, less
than 70% of the insects succumbed to the disease, and it could not be
transmitted to the next generation (BIRD 1955). In a different trial, tests
against the African armyworm with its respective NPV also proved ineffective (BROWN and SWAINE 1965).

d) Toxicity and pathogenicity

As stated by HEIMPEL (1971): "(the) data, collectively, indicate that
the insect viruses are the safest of all microbial agents proposed for in-

sect control and are infinitely safer for other life forms than chemical agents."

Although the evidence does support the view that most insect viruses are safe for use, in a few cases there have been exceptions. The following material merely outlines rather than details the nature of these pathogens.

1. Infections.—For beneficial insects, few studies have been done on the virulence of some viruses to cross-infect other hosts. In one investigation, however, 10^{10} polyhedra from cabbage loopers, corn earworms, greater wax moths, and redbanded leaf rollers were fed to bee colonies with no ill effects (CANTWELL et al. 1966).

That cross-infection may occur is shown by studies on the Tipula iridescent virus (TIV) which could be cross-inoculated into flies, butterflies, and moths (SMITH et al. 1961). The significance of such tests is doubtful, as laboratory conditions do not mimic those of the field while administration of the virus by injections may bypass inherent barriers. As an example, bee-paralysis virus is harmless when 10^{10} particles are fed to bees, yet 10^2 particles will cause disease if injected (BAILEY and GIBBS 1964).

When cross-infection does occur it may not always be a negative character. For several insects infesting a field, a narrow-spectrum virus would simply allow the unaffected species to proliferate. In a few instances, NPV has been shown to infect more than one pest (VAIL et al. 1978), enhancing its usefulness.

Concerning vertebrates, it was previously believed that either viruses attacked a narrow host range of insects or, as in the case of arboviruses, were pathogenic only to the vertebrates. An exception appears to be the Nodamura virus. Not only does it fatally infect honeybees and the greater wax moth, but it can also grow in mammalian cell cultures and kill suckling mice (BAILEY et al. 1975). Therefore, not all insect viruses are infinitely safer than chemical agents for use against pests.

2. Toxicity.—IGNOFFO and HEIMPEL (1965) investigated the effects of the NPV of the cotton bollworm, Heliothis zea, and found the following:

1. Mice were exposed to 500 mg of virus powder and guinea pigs to 100 mg for 2 hr each day for 2 mon. No signs of disease were detected and all animals showed normal weight gain.

2. Shaved guinea pigs were injected intradermally on the back with polyhedra, insect debris plus polyhedra, and insect debris with polyhedra removed. Only the first group of animals was free from welts indicating the lack of antigenicity of the virus.

3. Guinea pigs were injected intravenously with freed virus and the polyhedra protein. During the 2 mon investigation, all animals were healthy with normal weight gain.

4. White mice were force-fed polyhedra in water. During the 2-mon of study, none showed signs of illness.

5. Newly weaned white mice were injected intracerebrally with large doses of freed virus. All animals were normal after 2 mon.

Other investigations have been acute and chronic feeding tests, observations on rodent progeny for abnormal effects, assessment of carcinogenic potential, and feeding experiments to a variety of fish, shellfish,
and birds, as well as the effects on mammalian cell cultures. All were
negative (HEIMPEL 1971).

Feeding experiments have also been reported on human volunteers.
In one case, 10 men and women were each fed 5.8 × 10⁹ polyhedra
over 5 days while the control group of 6 men and women were fed insect
protein separated from the same viral preparation. No significant effects
were found after 2 yr (HEIMPEL and BUCHANAN 1967).

Although the above deals with the nuclear polyhedrosis virus from
the cotton bollworm, similar results have been found with those from
the cabbage looper (IGNOFFO and HEIMPEL 1965). The extremely low
toxicity shows that at least these viruses are safe for commercial uses.

VI. Bacteria

a) Background

Correlations between bacteria and diseases in insects began in 1870
with Pasteur's study of flacherie in silkworms. Since then, bacteria as
microbial control agents have received considerable attention due to
the availability of methods for identification and cultivation. Basically
two strategies are used in the study of pathogenicity (FAST 1973):

1. Isolating microorganisms from insect cadavers and then testing
these against healthy animals.

2. Studying the bacterial populations of healthy insects and then
determining what conditions will cause disease (CHARPENTIER *et al.*
1978).

By far the first approach has the greater chance of success for, of
the hundreds of bacteria in the gut flora, only a few are significant pathogens under abnormal conditions, conditions which may not be duplicated
naturally in the field (FAST 1973).

b) Mode of infection

Most of the bacteria that are ingested with the foliage, debris, or soil
do not usually affect the feeding insect, nor are they of an exotic classification. Those within the midgut of turnip moths, for example, are
*Streptococcus faecalis, Citrobacter freundii, Klebsiella pneumoniae, Pro-
teus vulgaris, Proteus mirabilis,* and *Escherichia coli* (CHARPENTIER *et al.*
1978)—strains which are frequently found commensal to people. When
the insect is subjected to stress such as infections with other pathogens,
lack of food, or food contaminated with fecal material, then these agents
may invade the animal to kill it by bacteremia (BUCHER 1967).

A few strains, notably the spore-formers, however, have evolved a

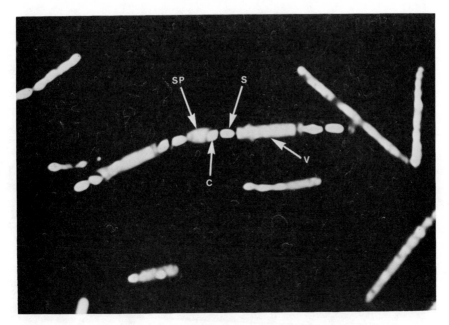

Fig. 3. Stages of development in the sporulating cells of *Bacillus thuringiensis* (*v*)
vegetative cell (*sp*) sporangium (*s*) spore (*c*) crystal (× 4500).

(Photograph by Dr. T. A. Angus, Great Lakes Research Centre and Insect Pathology
Research Institute, Sault Ste. Marie, Ontario.)

more complex mechanism of infection. As shown in Figure 3, upon further development, the vegetative cell of *Bacillus thuringiensis* becomes
a sporangium, producing a spore and a parasporal crystal. The diamond-
shaped crystal (see Fig. 4), is a collection of protoxins. When a susceptible host ingests a leaf contaminated with these, initially the crystal
dissolves in the alkaline medium of the insect's gut (FAUST *et al.* 1967).
The digestive enzymes of the insect then degrade these proteins, releasing polypeptide fragments (LECADET and DEDONDER 1967) which
cause a change in permeability of the epithelial cells (FAST and ANGUS
1965). Within 30 min these cells burst, greatly distorting the gut wall.
The pH drops, the spores germinate, and then invade the hemocoel
where they multiply rapidly and destroy the tissues. Following death,
the host disintegrates, releasing the bacteria to contaminate another leaf,
hence completing the cycle (FALCON 1971, FAST 1973). Death of the
insect can result from ingestion of spores, crystals, or exotoxins, although
a collection of these is more effective (BURGERJON and MARTOURET 1971).
    Other spore-formers infect their hosts simply by the germination of
their spores followed by a bacteremia. *Bacillus popilliae* can only grow
within the white grubs of the Japanese beetle (DUTKY 1963), while

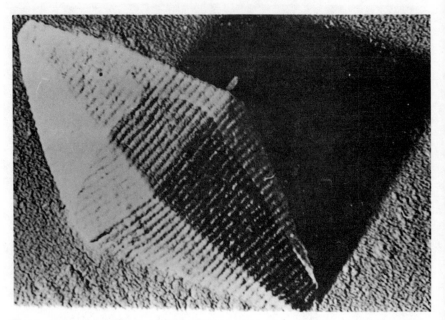

Fig. 4. A scanning electron micrograph of a parasporal protein crystal from *Bacillus thuringiensis* (× 250,000).

(Photograph by Dr. C. Hannay, Great Lakes Research Centre and Insect Pathology Research Institute, Sault Ste. Marie, Ontario.)

B. cereus is a common soil saprophyte with a wide host range (HEIMPEL and ANGUS 1963). *Clostridia* species, which are obligate insect pathogens, differ from the above in that they do not invade the hemocoel but, instead, only multiply within the gut (FALCON 1971).

c) Field trials

B. thuringiensis and *B. popilliae* are perhaps the best examples of how microbial agents can be used as insecticides of commercial value. Hundreds of tons of spore products are manufactured each year in five countries and production is rising annually. *B. thuringiensis* is registered for a variety of uses (see Table I) and is regarded by the U.S. Department of Agriculture as both safe and effective in field trials against the registered pest species (FALCON 1971). A few trademark names for this species are Biotrol (U.S.A.), Biospar (Germany), Sporeine (France), Baktukal (Yugoslavia), Bathurin (Czechoslovakia), and Entobacterin 3 (U.S.S.R.). *B. popilliae* is sold under the names Doom and Japidemic in the United States (VAN DER GEEST *a*nd VAN DER LAAN 1971). Estimated costs of the material for the user lie between $1.75 to $7.50/A,

Table I. *Registered uses for* Bacillus thuringiensis *products in Canada.*[a]

Pest	Protected plant
Agricultural application	
Cabbage looper	Broccoli, brussels sprouts, cabbage, cauliflower, celery, kale, lettuce, mustard greens, potatoes, spinach, tomatoes
Diamond moth and imported cabbageworm	Broccoli, brussels sprouts, cabbage, cauliflower
Essex Skipper	Timothy mixtures and other forage grasses
Hornworms	Tobacco, tomatoes
Rangeland caterpillar	Rangeland
Greenhouse use	
Cabbage looper	Chrysanthemum, tomatoes
Hornworms	Tomatoes
Omnivorous leafroller	Roses
Tree application	
Bagworm	Ornamental and shade trees in farm woodlots, municipal parks, and rights-of-way
Elm spanworm	
Fall webworm	
Gypsy moth	
Spring and fall cankerworm	
Tent caterpillar	
Spruce budworm	Balsam fir and spruce trees in Christmas tree plantations, farm woodlots, municipal parks, and rights-of-way

[a] From information taken in part through the kindness of Sandoz, Inc., Crop Protection, 480 Camino Del Rio South, San Diego, CA 92108.

making them a competitive alternative to chemical insecticides (FALCON 1971).

Since the 1940s, there have been numerous field trials dealing with varieties of these two bacteria, proving their effectiveness (FALCON 1971). As one example, in the initial trial of the "milky disease" against the Japanese beetle in 1939, the number of grubs on a golf course was reduced from 20 to 60 grubs/sq ft of turf to 1 to 3 grubs/sq ft. Nine years later the number was still at this level without further applications (DUTKY 1963). As these bacteria grow poorly outside their respective hosts, this persistence in the environment can be explained in part by the resistance of their spores. In 1964, IGNOFFO (as reported by FALCON 1971) found that dry preparations of B. thuringiensis were viable after 200 days at 10 to 30°C, while samples in water were viable after 500 days at 10°C. Under exposure to solar radiation, however, particularly at the 400 nm wavelength region, the spores were quickly inactivated (CANTWELL 1967, FRYE et al. 1973, GRIEGO and SPENCE 1978). It seems likely, therefore, that the persistence may also be due to passage and replication in susceptible hosts.

d) Toxicity and pathogenicity

1. Infections.—For nonspore formers, bacteria that were found to take advantage of the stressed insect have also been noted in human infections. *Pseudomonas* species have been reported in those suffering from burns, endocarditis, pneumonia, urologic problems, and scratched corneae (Flick and Cuff 1974, Davis and Chandler 1974), while *Serratia marcescens* has caused severe epidemics in nurseries (Simberkoff et al. 1976). The use of these for biological control is therefore unlikely. Even less likely are the uses of known pathogens such as *Salmonella*, despite some sensitivity of insects such as the greater wax moth (Kurstak and Vega 1967).

For *Bacillus thuringiensis* or *B. popilliae*, there has been no evidence of cross-infectivity to anything outside their specific target hosts. Such limitations for disease are understandable. For the former bacterium, as an example, a strongly alkaline gut pH of 9.0 to 10.5 (Heimpel and Angus 1960), specific proteolytic enzymes (Lecadet and Martouret 1967), and suitable tissue receptor sites (White as reported by Cooksey 1971, Sutter and Raun 1967) are required. In an experiment that shows the usefulness of this specificity, 3×10^6 spores/g of brood comb were given as small blocks to a bee colony. The bees used this material to build their hive with no ill effects. The wax moth, which was the target insect for the impregnated spores, was still susceptible and, hence, the bee colony was protected from this pest (Burges and Bailey 1968).

2. Toxicity.—As *B. thuringiensis* can be grown on artificial media, Brown et al. (1958) tested blood-dextrose medium to see if it would acquire pathogenicity for mice. Results were negative. In further studies by Fisher and Rosner (1959) their results were given as follows:

1. Mice were injected intraperitoneally with a spore suspension and 6 hr later blood was drawn by cardiac puncture and injected intraperitoneally into a second group. No mice died but the second group did show severe irritation of the peritonea.

2. Guinea pigs were injected intraperitoneally with vegetative cells of broth and slant cultures. None died from the former but 7/10 did die from the latter.

3. Mice were exposed 4 times with gram quantities of the spore powder dispersed by a powder blower in a confined 30-cm^2 chamber. There were no changes in feeding habits or in pathological examinations.

4. Guinea pigs had spores injected subcutaneously and applied topically. Only a slight erythema was noticed for the former, and no effects were observed with challenges 2 wk later.

5. Eighteen people each ingested a gram of spore preparations (3×10^9 viable spores/gm) each day for 5 days. Also, 5 other volunteers each inhaled 100 mg of this powder daily for 5 days. All subjects remained healthy over 5 wk of observation.

6. Acute and chronic feeding tests were done on chicks, laying hens, young swine and hogs, fish, and honeybees. All results were negative.

7. The data also included the records of 8 employees exposed for a total of 25 hr in the preparation of these spore samples. All personnel remained in good health.

These studies indicated that, at least with injections, *Bacillus thuringiensis* may prove pathogenic to vertebrates. However, as massive doses were used by a route out of the ordinary for exposure, the significance of these results is not known. In contrast, the usual routes of ingestion and inhalation showed no effect.

For *B. popilliae*, investigations show similar results with one humorous precedent. In Maryland, a woman who owned a farm and had used the "milky disease" lost some chickens. To collect her losses from the county commissioners, a hearing was held and research scientists were called to testify. When Dr. George Langford of the University of Maryland was called to the stand, he stated that he believed the bacterium was harmless to vertebrates, then swallowed a heaping spoonful of the spore preparation, washing it down with a glass of water. The magistrate immediately suspended the case pending the outcome of Dr. Langford. Years later, he was still alive and healthy and had gone on to become the Maryland State Entomologist (as reported by HEIMPEL 1971).

Acknowledgments

The author is grateful for the assistance given by the researchers at the Great Lakes Forest Research Centre and Insect Pathology Research Institute, Sault Ste. Marie, Ontario, Canada, notably J. C. Cunningham. The manuscript also benefited greatly from the comments and criticisms given by W. Kalow, Department of Pharmacology, University of Toronto, Toronto, Ontario, Canada.

Summary

This review describes several microbial agents as being alternatives to chemical insecticides. Of low toxicity and high specificity, they have also other advantages. By being a natural part of the environment (*i.e.*, some microbes appear to have evolved with their respective hosts), their application to a field does not upset the ecological balances at work. Instead, they merely correct an already existing imbalance, returning the numbers of a pest back to a tolerable level. They also have the advantage in that they are living pesticides, in some cases protecting a field for years after application by infecting and spreading from an insect to its progeny.

Of the microbial agents reviewed, certain viruses and bacteria have already proved successful. The nuclear polyhedrosis viruses (NPV) have been useful in controlling sawflies, butterflies and moths, while *Bacillus thuringiensis* and *B. popilliae* are commercially available for use against

92 J. W. CHERWONOGRODZKY

several pests. This does not mean that all microbial agents are equally beneficial. Fungi and protozoa, although also a natural part of the environment, are less specific and act slowly. As for rickettsiae, it seems unlikely that these will ever be used for controlling insects as this group appears more pathogenic for vertebrates. Only by understanding the potentials and limitations of these agents, therefore, can one best decide if their use against a pest has merit. Such information is frequently lacking and further investigations should be strongly encouraged.

References

ADAMS, M. W., A. H. ELLINGBOE, and E. C. ROSSMAN: Biological uniformity and disease epidemics. Bioscience. **21**, 1067 (1971).
ALEXOPOULOS, C. J.: Introductory mycology, 2 ed., pp. 6–8, 412–417. London-New York-Sydney: Wiley (1962).
ARISON, R. N., J. A. CASSARO, and M. P. PRUSS: Studies on a murine ascites producing agent and its effect on tumor development. Cancer Res. **26**, 1915 (1966).
BAILEY, L.: The safety of pest-insect pathogens for beneficial insects. In H. D. Burges and N. W. Hussey (eds.): Microbial control of insects and mites, pp. 491–503. London-New York: Academic Press (1971).
——, and A. J. GIBBS: Acute infection of bees with paralysis virus. J. Insect Pathol. **6**, 395 (1964).
——, J. F. E. NEWMAN, and J. S. PORTERFIELD: The multiplication of Nodamura virus in insect and mammalian cell cultures. J. Gen. Virol. **26**, 15 (1975).
BELL, J. V.: Mycoses. In G. Cantwell (ed.): Insect diseases, vol. 1, pp. 185–236. New York: Marcel Dekker (1974).
BIRD, F. T.: Virus diseases of sawflies. Can. Entomologist **87**, 124 (1955).
——, and J. M. BURK: Artificially disseminated virus as a factor controlling the European spruce sawfly, *Diprion hercyniae* (Htg.), in the absence of introduced parasites. Can. Entomologist **93**, 228 (1961).
BRAS, G., C. C. GORDON, C. W. EMMONS, K. M. PENDEGAST, and M. SUGAR: A case of phycomycosis observed in Jamaica, infection with *Entomophthora coronata*. Amer. J. Trop. Med. Hyg. **14**, 141 (1965).
BRIDGES, C. H., W. M. ROMAINE, and C. W. EMMENS: Phycomycosis of horse caused by *Entomophthora coronata*. J. Amer. Vet. Med Assoc. **140**, 673 (1962).
BROWN, E. R., M. D. MOODY, E. L. TREECE, and C. W. SMITH: Differential diagnosis of *Bacillus cereus*, *Bacillus anthracis*, and *Bacillus cereus* var. Mycoides. J. Bacteriol. **75**, 499 (1958).
BROWN, E. S., and G. SWAINE: Virus disease of the African armyworm, *Spodoptera exempta* (Wlk.). Bull. Entomol. Res. **56**, 95 (1965).
BUCHER, C. E.: Pathogens of tobacco and tomato hornworms. J. Invertebrate Pathol. **9**, 82 (1967).
BURGERJON, A., and D. MARTOURET: Determination and significance of host spectrum of *Bacillus thuringiensis*. In H. D. Burges and N. W. Hussey (eds.): Microbial control of insects and mites, p. 265. London-New York: Academic Press (1971).
BURGES, H. D., and L. BAILEY: Control of the greater and lesser wax moths (*Galleria mellonella* and *Achroia grisella*) with *Bacillus thuringiensis*. J. Invert. Pathol. **11**, 184 (1968).
——, and N. W. HUSSEY: Introduction. In H. D. Burges and N. W. Hussey (eds.): Microbial control of insects and mites, pp. 1–11. London-New York: Academic Press (1971 a).
—— —— Past achievements and future prospects. In H. D. Burges and N. W. Hussey (eds.): Microbial control of insects and mites, pp. 687–709. London-New York: Academic Press (1971 b).

CANTWELL, G. E.: Inactivation of biological insecticides by irradiation, J. Invert. Pathol. 9, 138 (1967).
——, D. A. KNOX, T. LEHNERT, and A. S. MICHAEL: Mortality of the honeybee, *Apis mellifera*, in colonies treated with certain biological insecticides. J. Invert. Pathol. 8, 228 (1966).
CHARPENTIER, R., B. CHARPENTIER, and O. ZETHER: The bacterial flora of the midgut of two Danish populations of healthy fifth instar larvae of the turnip moth, *Scotia segetum*. J. Invert. Pathol. 32, 52 (1978).
COOKSEY, K. E.: The protein crystal toxin of *Bacillus thuringiensis*: Biochemistry and mode of action. In H. D. Burges and N. W. Hussey (eds.): Microbial control of insects and mites, p. 265. London-New York: Academic Press (1971).
CROW, J. F.: Genetics of insect resistance to chemicals. Ann. Rev. Entomol. 2, 227 (1957).
CUNNINGHAM, J. C.: The use of viruses in the control of forest insect pests. In: Alternatives in forest insect control, symopsium proceedings O-P-2, pp. 46–52. *Great Lakes Forest Research Centre and Insect Pathology Research Institute*, Sault Ste. Marie, Ontario (1973).
DAVID, W. A. L., B. O. C. GARDINER, and M. WOOLNER: The effect of sunlight on a purified granulosis virus of *Pieris brassicae* applied to cabbage leaves. J. Invert. Pathol. 11, 496 (1968).
DAVIS, B. D., R. DULBECCO, H. N. EISEN, H. S. GINSBERG, and W. B. WOOD, JR.: Microbiology, pp. 928–959. Evanston-London-New York: Harper & Row (1970).
DAVIS, S. D., and J. W. CHANDLER: Experimental keratitis due to *Pseudomonas aeruginosa*: model for evaluation of antimicrobial drugs. Antimicrob. Agents Chemother. 8, 350 (1975).
DIEUZEIDE, R.: Les champignons entomophytes du genre Beauveria Vuillemin. Contribution à l'étude de *Beauveria effusa* Vuill. parasite du Doryphore. Ann. épiphyt. 11, 185 (1925).
DUSTMAN, E. H., and L. F. STICKEL: The occurrence and significance of pesticide residues in wild animals. Ann. N. Y. Acad. Sci. 160, 162 (1969).
DUTKY, S. R.: The milky diseases. In E. A. Steinhaus (ed.): Insect pathology, an advanced treatise, vol. 2, p. 97. London-New York: Academic Press (1963).
EASWARAMOORTHY, S., and S. JAYARAJ: Effectiveness of the white halo fungus, *Cephalosporium lecanii*, against field populations of coffee green bug, *Coccus viridis*. J. Invert. Pathol. 32, 88 (1978).
FALCON, L. A.: Use of bacteria for microbial control. In H. D. Burges and N. W. Hussey (ed.): Microbial control of insects and mites, pp. 67–90. London-New York. Academic Press (1971).
FAST, P. G.: The use of bacterial insecticides in forest insect pest control. In: Alternatives in forest insect control, symposium proceedings O-P-2, pp. 42–43. *Great Lakes Research Centre and Insect Pathology Research Institute*, Sault Ste. Marie, Ontario (1973).
——, and T. A. ANGUS: Effects of parasporal inclusions of *Bacillus thuringiensis* var. sotto Ishiwata on the permeability of the gut wall of *Bombyx mori* (Linnaeus) larvae. J. Invert. Pathol. 7, 29 (1965).
FAUST, R. M., J. R. ADAMS, and A. M. HEIMPEL: Dissolution of the toxic parasporal crystals from *Bacillus thuringiensis* var. *pacificus* by the gut secretions of the silkworm, *Bombyx mori*. J. Invert. Pathol. 9, 488 (1967).
FISHER, R. A., and L. ROSNER: Toxicology of the microbial insecticide, Thuricide J. Agr. Food Chem. 7, 686 (1959).
FLICK, M. R., and L. E. CUFF: *Pseudomonas* bacteremia—A review of 108 cases. Amer. J. Med. 60, 501 (1976).
FRYE, R. D., C. G. SCHOLL, E. W. SCHOLZ, and B. R. FUNKE: Effect of weather on a microbial insecticide. J. Invert. Pathol. 22, 50 (1973).
GIROUD, P., N. DUMAS, and B. HURPIN: Essais d'adaptation à la souris blanche de la rickettsie, agent de la maladie bleue de *Melolontha melolontha* L.: voie pulmonaire et voie buccale. Compt. Rend. Acad. Sci. 247, 2499 (1958).

GOLDBLATT, L. A.: Introduction. In L. A. Goldblatt (ed.): Aflatoxin, scientific background, control, and implications, pp. 1–11. London-New York: Academic Press (1969).

GRIEGO, V. M., and K. D. SPENCE: Inactivation of Bacillus thuringiensis spores by ultraviolet and visible light. Applied Environ. Microbiol. 35, 906 (1978).

HALEY, A. J.: Further observations on Glugea hertwigi Weissenberg 1911, 1913 (Microspordia) in fresh water smelt in New Hampshire. J. Parasitol. 40, 482 (1954).

HEIMPEL, A. M.: Safety of insect pathogens for man and vertebrates. In H. D. Burges and N. W. Hussey (eds.): Microbial control of insects and mites, pp. 469–487. London-New York: Academic Press (1971).

——, and T. A. ANGUS: Diseases caused by sporeforming bacteria. In E. A. Steinhaus (ed.): Insect pathology, an advanced treatise, vol. 2, pp. 40–43. London-New York: Academic Press (1963).

——, and L. K. BUCHANAN: Human feeding tests using a nuclear polyhedrosis virus of Heliothis zea. J. Invert. Pathol. 9, 55 (1967).

HENRY, J. E., E. A. OMA, and J. A. ONSAGER: Relative effectiveness of ULV spray applications of spores of Nosema locustae against grasshoppers. J. Econ. Entomol. 71, 629 (1978).

HOWSE, G. M.: The use of insecticides in the control of forest insect pests. In: Alternatives in forest insect control, symposium proceedings O-P-2, pp. 5–9. Great Lakes Forest Research Centre and Insect Pathology Research Institute, Sault Ste. Marie, Ontario (1973).

HUFFAKER, C. B., F. J. SIMMONDS, and J. E. LAING: The theoretical and empirical basis of biological control. In C. B. Huffaker and P. S. Messenger (eds.): Theory and practice of biological control, pp. 42–67. London-San Francisco-London: Academic Press (1976).

IGNOFFO, C. M.: The nuclear-polyhedrosis virus of Heliothis zea (Boddie) and Heliothis virescens (Fabricius). IV. Bioassay of virus activity. J. Invert. Pathol. 7, 315 (1965).

——, and A. M. HEIMPEL: The nuclear-polyhedrosis virus of Heliothis zea (Boddie) and Heliothis virescens (Fabricius). V. Toxicity—pathogenicity of virus to white mice and guinea pigs. J. Invert. Pathol. 7, 329 (1965).

KABATA, Z.: On two little-known microspordia of marine fish. Parasitol. 49, 309 (1959).

KELLEN, W. R., H. C. CHAPMAN, T. B. CLARK, and J. E. LINDEGREN: Transovarian transmission of some Thelohania (Nosematidea: Microsporidia) in mosquitoes of California and Louisiana. J. Invert. Pathol. 8, 355 (1966).

KOIDSUMI, K.: Antifungal action of cuticular lipids in insects. J. Insect Physiol. 1, 40 (1957).

KRAYBILL, H. F., and R. E. SHAPIRO: Implications of fungal toxicity to human health. In L. A. Goldblatt (ed.): Aflatoxin—scientific background, control, and implications, pp. 401–441. London-New York: Academic Press (1969).

KREIG, A.: Untersuchungen für Wirbeltierpathogenitat und zum serologischen Nachweis der Rickettsia melolonthae im Arthropod-Wirt. Naturwissensch. 42, 609 (1953).

—— Virus—und Rickettsien-Erkankungen bei Lamellicorniern. Z. Parasitenk. 21, 309 (1962).

—— Rickettsiae and rickettsioses. In E. A. Steinhaus (ed.): Insect pathology, an advanced treatise, vol. 1, p. 605. London-New York: Academic Press (1963).

—— Possible use of rickettsiae for microbial control of insects. In H. D. Burges and and N. W. Hussey (eds.): Microbial control of insects and mites, pp. 173–178. London-New York: Academic Press (1971).

KRYWIENCZYK, J., and G. H. BERGOLD: Serological studies of inclusion body proteins by antigen and diffusion technique. J. Insect Pathol. 3, 15 (1961).

KURSTAK, E., and C. E. VEGA: Infection bactérienne à Salmonella typhimurium chez un invertebré, Galleria mellonella L. Can. J. Microbiol. 14, 233 (1967).

LAIRD, M.: Microbial control of arthropods of medical importance. In H. D. Burges and N. K. Hussey (eds.): Microbial control of insects and mites, pp. 387–403. London-New York: Academic Press (1971).

LECADET, M. M., and R. DEDONDER: Enzymatic hydrolysis of the crystals of *Bacillus thuringiensis* by the proteases of *Pieris brassicae*. I. Preparation and fractionation of the lysates. J. Invert. Pathol. 9, 310 (1967).

——, and D. MARTOURET: Enzymatic hydrolysis of the crystals of *Bacillus thuringiensis* by the proteases of *Pieris brassicae*. J. Invert. Pathol. 9, 322 (1967).

LICHTENSTEIN, E. P.: Persistence and degradation of pesticides in the environment. In: Scientific aspects of pest control, publication number 1402, *National Academy of Science, National Research Council,* Washington, D. C. (1966).

LYNN, D. E., and W. F. HINK: Infection of synchronized TN-368 cell cultures with alfalfa looper polyhedrosis virus. J. Invert. Pathol. 32, 1 (1978).

MACLEOD, D. M.: Investigations on the genera *Beauveria* Vuill. and *Tritirachium limber.* Can. J. Bot. 32, 818 (1954).

MARTIGNONI, M. E., and P. SCHMID: Studies on the resistance to virus infections in natural populations of Lepidoptera. J. Insect Pathol. 3, 62 (1961).

MATSUBAYASHI, H., T. KOIKE, I. MIKAYA, H. TAKEI, and S. HAGIWARA: A case of encephalitozoa-like body infection in man. Arch. Pathol. 67, 181 (1959).

McLAUGHLIN, R. E.: Use of protozoans for microbial control of insects. In H. D. Burges and N. W. Hussey (eds.): Microbial control of insects and mites, pp. 151–170. London-New York: Academic Press (1971).

——, T. C. CLEVELAND, R. J. DAUM, and M. R. BELL: Development of the bait principle for boll weevil control. IV. Field tests with a bait containing a feeding stimulant and the sporozoans, *Glugea gasti* and *Mattesia grandis.* J. Invert. Pathol. 13, 429 (1969).

MOORE, H. H., A. M. HAMMOND, and G. C. LLEWELLYN: Chemosterilant and insecticidal activity of mixed aflatoxins against *Anthonomus grandis* (Coleoptera). J. Invert. Pathol. 31, 365 (1978).

NIKLAS, O. F.: Zur Temperaturabhängigkeit der Vertikaibewegungen Rickettsiose-Kranker Maikäfer-Engerlinge. Anz. Schädlingskunde 30, 113 (1957).

—— Standorteinflüsse und naturliche Feinde als Begrenzungsfaktoren von *Melolontha*—Larvenpopulationen eines Waldgebietes (Forstamt Lorsch, Hessen) (Coleoptera: Scarabaeidae). Mitt. Biol. Bundesanstalt Land—u. Forstwirtsch. Berlin-Dahlem 101, 60 (1960).

PATTERSON, D. S. P.: Metabolism as a factor in determining the toxic action of the aflatoxins in different animal species. Food Cosmet. Toxicol. 11, 287 (1973).

PETRI, M.: The occurrence of *Nosema cuniculi* (Encephalitozoon cuniculi) in the cells of transplantable, malignant ascites tumours and its effect upon tumour and host. Acta Pathol. Microbiol. Scand. 66, 13 (1966).

PRASERTPHON, S., and Y. TANADA: The formation and circulation, in *Galleria*, of hyphal bodies of Entomophthoraceous fungi. J. Invert. Pathol. 11, 260 (1968).

PRIESTER, T. M., and G. P. GEORGHIOU: Inheritance of resistance to permethrin in *Culex pipens quinquefasciatus.* J. Econ. Entomol. 72, 124 (1979).

RESTROPO, A. M., D. L. GREER, M. V. ROBLEDO, C. G. DIAZ, R. M. LOPEZ, and C. R. BRAVO: Subcutaneous phycomycosis: Report of the first case observed in Columbia, South America. Amer. J. Trop. Med. Hyg. 16, 34 (1967).

ROBERTS, D. W., and W. G. YENDOL: Use of fungi for microbial control of insects. In H. D. Burges and N. W. Hussey (eds.): Microbial control of insects and mites, pp. 125–145. London-New York: Academic Press (1971).

Secretary's Commission on Pesticides: Report of the Secretary's Commission on Pesticides and their relationship to environmental health. *U.S. Department of Health, Education, and Welfare,* Washington, D. C.: U. S. Government Printing Office (1969).

SIMBERKOFF, M. S., I. RICUPERO, and J. J. RAHAL, JR.: Host resistance to *Serratia marcescens* infections: Serum bactericidal activity and phagocytosis by normal blood leukocytes. J. Lab. Clin. Med. 87, 206 (1976).

SMIRNOFF, W. A.: Effects of volatile substances released by foilage of various plants on the Entomopathogenic *Bacillus cereus* group. J. Invert. Pathol. 11, 513 (1968).

SMITH, K. M., G. J. HILLS, and C. F. RIVER: Studies on the cross-inoculation of the *Tipula* iridescent virus. Virology 13, 233 (1961).

—— ——, F. MUNGER, and J. E. GILMORE: A suspected virus disease of the citrus red mite, *Panonychus citri* (McG.) Nature 184, 70 (1959).

SMITH, R. F., and R. VAN DEN BOSCH: Integrated control. In W. W. Kilgore and R. L. Doutt (eds.): Pest control, pp. 295–337. London-New York: Academic Press (1967).

STAIRS, G. R.: Use of viruses for microbial control of insects. In H. D. Burges and N. W. Hussey (eds.): Microbial control of insects and mites, pp. 97–122. London-New York: Academic Press (1971).

STANIER, R. Y.: The microbial world, 3 ed., pp. 109–115. Englewood Cliffs: Prentice Hall (1970).

STICKNEY, A. P.: A previously unreported peridinian parasite in the eggs of the Northern shrimp, *Pandulus borealis*. J. Invert. Pathol. 32, 212 (1978).

STUART-HARRIS, C. H.: The rickettsial diseases. In S. Bedson, A. W. Downie, F. O. MacCallum, and C. H. Stuart-Harris (eds.): Virus and rickettsial diseases of man, pp. 410–449. London: Edward Arnold (1967).

STUNKARD, H. W., and F. E. LUX: A microsporidian infection of the digestive tract of the winter flounder, *Pseudopleuronectes americanus*. Biol. Bull. 129, 371 (1965).

SUTTER, G. R., and E. S. RAUN: Histopathology of European-corn-borer larvae treated with *Bacillus thuringiensis*. J. Invert. Pathol. 9, 90 (1967).

TYRRELL, D., and D. M. MacLEOD: Alternatives in the control of forest insects: Fungi. In: Alternatives in forest insect control, symposium proceedings O-P-2, pp. 53–58. *Great Lakes Forest Research Centre and Insect Pathology Research Institute*, Sault Ste. Marie, Ontario (1973).

VAIL, P. V., T. J. HENNEBERRY, and M. R. BELL: Comparative susceptibility of *Heliothis virescens* and *H. zea* to the nuclear polyhedrosis virus isolated from *Autographa californica*. J. Econ. Entomol. 70, 727 (1978).

——, D. L. JAY, and W. F. HINK: Replication and infectivity of the nuclear polyhedrosis virus of the alfalfa looper, *Autographa californica*, produced in cells grown *in vitro*. J. Invert. Pathol. 22, 231 (1973).

VAN DER GEEST, L. P. S., and P. A. VAN DER LAAN: Sources of special materials. In H. D. Burges and N. W. Hussey (eds.): Microbial control of insects and mites, pp. 744–745. London-New York: Academic Press (1971).

VEEN, K. H.: Oral infection of second-instar nymphs of *Schistocerca gregaria* by *Metarrhizium anisopliae*. J. Invert. Pathol. 8, 254 (1966).

WEISER, J.: On the taxonomic position of the genus *Encephalitozoon* Levaditi, Nicolau and Schoen, 1923 (protozoa, microsporidia). Parasitol. 54, 749 (1964).

——, and J. VEBER: Die Mikrosporidie *Thelohania hyphantriae* im weissen Bärenspinner und anderen Mitgliedern seiner Biocoenose. Z. Angew. Entomol. 40, 55 (1956).

WHITTEN, C. J.: Inheritance of methyl parathion resistance in tobacco budworm larvae. J. Econ. Entomol. 71, 971 (1978).

WOGAN, G. N.: Biochemical responses to aflatoxins. Cancer Res. 28, 2282 (1968).

YENDOL, W. G., and J. D. PASCHKE: Pathology of an *Entomophthora* infection in the eastern subterranean termite *Reticulifermes flavipes* (Kollar). J. Invert. Pathol. 7, 414 (1965).

Manuscript received January 29, 1980; accepted February 16, 1980.

Opening session of the US-ROC Seminar on Environmental Problems Associated With Pesticide Usage in the Intensive Agriculture System, April 7–17, 1979, Taipei, Taiwan

By

Dr. S. S. Shu, Chairman of the National Science Council

First of all I would like to extend to you our warmest welcome for attending this seminar, particularly for those who have traveled long distances such as from the United States.

A great many seminars and international forums have been held to discuss the problem of pesticides. They had intended in many instances to discuss the basic science of pesticide chemistry, usage, side effects etc. This particular seminar has been designed to address itself to the environmental problems associated with pesticide usage in the intensive agricultural system.

Pesticide usage in Taiwan is massive both in the total volume of pesticides used and in the diversity of chemicals used. So far, about 371 individual chemical compounds are registered and the total annual volume of pesticides now used is equivalent to 3.1 kg active ingredients per hectare of Taiwan's agricultural land. The rate of usage is among the highest in the world and is about twice that of the corn and soybean croplands of the U.S. midwest or of California's central valley. Moreover, Taiwan with its limited area, intensive cultivation, and very high population density is more vulnerable to nontarget effects than a large continental area. I sincerely believe that the constant exchanges of knowledge, experience, and ideas during this seminar and afterwards will help us to identify the most important problems associated with pesticide usage in this unique environmental setting. The results will not only benefit both of our respective countries but also the world, and in particular the southeast Asian countries.

This seminar will consist of four sessions in which papers will be presented on selected topics.* These four sessions are (1) monitoring

---

* Editor's note: Only the manuscripts from the United States' delegates are presented in the following pages; the manuscripts from the Taiwanese delegates were not submitted for publication, much to our collective regret.

Residue Reviews, Volume 76

pesticide residues in the environment and in food, (2) pesticide degradation, (3) chemistry, and (4) the effects on nontarget organisms. It appears to be an excellent program covering many subjects on pesticides and their residues. I wish that, through your deliberation, this seminar can achieve great successes. And I am anxious to have the meeting propose avenues of cooperative research on these subjects.

As you know, this is the first binational seminar jointly sponsored by the U.S. National Science Foundation and the National Science Council of the Republic of China after the rupture of diplomatic relations between our two governments on 1979's new year's day. While we were disappointed in the change of U.S. foreign policy which affects us, it is nevertheless a fortunate thing that both governments indicated that existing treaties between the two countries, including the science cooperation agreement, should be maintained. I would like to appeal to all of you here that, because of the rupture of diplomatic relations, it is all the more important to strengthen the traditional ties between our two peoples, especially the ones between our scientific communities. This is not only the wish of interested scientists of our two countries, it is also the command of our two peoples, as evidenced by the Taiwan Relations Act just adopted by the U.S. Congress and the fact that our Coordinating Council for North American Affairs already started operation in Washington in the place of our Embassy.

In addition to exchange of research ideas and information, this seminar is designed by its co-sponsors as a fountainhead for ideas of cooperative research between scientists of our two countries. On the part of the ROC National Science Council, I would like to reassure you that proposals for cooperative research generated by you will receive the best consideration for implementation here. I also have reason to believe that the National Science Foundation shares with us the same attitude. I would therefore suggest that, in your exploration of cooperative opportunities, you would not allow your choices and preferences to be beclouded by the vicissitudes of politics. My colleagues from the Republic of China, I know, will cherish the friendship of our American partners and will assume a "business as usual" approach. I sincerely hope our distinguished American guests will reciprocate in the same spirit. During the seminar you will have opportunities to tour our country. I trust our American friends will find that the cordial sentiment of every person in the Republic of China towards the American people is a durable and steadfast sentiment, and that we want to expand our cooperative ties with you in the realm of science, education, and culture, as well as in economic interactions.

Bioassay as a monitoring tool

By Gustave K. Kohn[*]

Contents

[*] Zoecon Corporation, 975 California Avenue, Palo Alto, CA 94306. Presented in part at the April 1979 US-ROC Cooperative Science Program seminar on "Environmental Problems Associated with Pesticide Usage in the Intensive Agricultural System," Taipei, Taiwan, Republic of China, as sponsored by the National Science Foundation (U.S.A.) and the National Science Council (R.O.C.).

I. Introduction

The application of technology to agriculture during the twentieth century has undoubtedly contributed to a vast increase in agricultural productivity and as a consequence an increase in the world's population. This, in turn, places even greater demand on improved scientific agricultural practice. One of the factors of that technology is the introduction of chemotherapeutic agents: chemicals for the control of insects, plant disease, weed management, plant growth regulation, rodent control, etc., and as well for soil improvement (fertilizers, amendments, etc.).[1]

With the acknowledgment, particularly but not exclusively in the last three decades of the toxicological and ecological consequences of these agents, methods for their quantitative estimation (including their metabolites) have been developed. Most of these methods have been physical and chemical (Gunther and Blinn 1955, Zweig 1963 et seq., Schechter and Hornstein 1957). The most sensitive and precise of the quantitative methods require the use of "high" technology—complex instrumentation, computerization, and sophisticated manipulation.

The development of chemotherapeutic agents generally requires a screening procedure, an assay method based upon organism interaction with the chemical under highly specific and controlled conditions. Wherever chemotherapeutic agents are utilized, for medicine, public health, sanitation, and agriculture, such bioassay methods have been developed.

Monitoring by bioassay for residues utilizes the same interactions of organisms and chemicals, and at its best requires the same specific and controlled conditions. This paper will be directed at practical methods that can be devised by the utilization of this reverse of the screening procedure. It will be concerned with useful procedures, some summarized in earlier publications (Busvine 1971, Sun 1963, Dewey 1958, Sun 1957), some unpublished, as well as some suggested herein.

It should be remembered that the vast majority of the chemotherapeutants have been developed by industrial corporations and their bioassay methods are quite frequently regarded as proprietary and are rarely published as such.

It is the intention of this brief review to provide examples rather than to be definitive and complete, for to do the latter requires a multivolume treatise. We will include assays for insecticides and insect growth regulators, herbicides and plant growth regulations, fungicides, and bactericides. We will treat residues in and on agricultural crops, stored grains, etc., soils, water, home usage and manufactured products, and animal waste products.

The outline for this paper divides the subject matter logically largely on the basis of toxicant function. Nature defies this logic. The assay methods frequently transcend boundary lines. Hence the section devoted

---

[1] Pesticide and other chemicals mentioned in text are identified in Table VII.

to algae assays will be found in the assay for herbicides section although, as will be seen, certain insecticides, fungicides, bactericides, antibiotics, and nematicides react with these algae.

II. Problems and limitations to the bioassay monitoring of residues

a) Bound and conjugated residues

The application of a pesticide to a crop or to the soil can result in residues with varying degrees of availability of the toxicant. The extreme example of this difference of availability is the adsorption of paraquat (RILEY et al. 1976) on montmorillonite-containing solids with agricultural and industrial effluents. This difference is reflected in a 0% availability as measured by bioassay (MATSAMURA 1975, ETO 1974, MENN et al. 1957) to 100% availability by chemical assay. In this case, 5 hr digestion with 18 N H_2SO_4 is required to break down the montmorillonite crystal lattice for the release of the chemically relatively inert paraquat. The bioassay provides insight to an important aspect of the agricultural hazard involved while the chemical assay more or less accurately describes the true residue in quantitative terms. This residue is present as chemically determined but is so tightly bound as to permit the germination of seeds and the growth of seedlings.

This example is cited because one of the sources of manufacture of diquaternary herbicides is here in Taiwan. Bound and conjugated residues have been the subject of a recent American Chemical Society symposium (KAUFMAN et al. 1976).

b) Systemic pesticides or metabolites

Careful planning is required for the bioassay of a systemic insecticide, fungicide, and herbicide as well as systemic metabolites. Obviously a surface wash and test of the concentrated washings will be totally inadequate. Special procedures using sucking or rasping insects have been devised for direct tests of plant tissue. BUSVINE (1971) provided procedures adequate for the bioassay of many systemics.

c) Temperature effects

BUSVINE (1971) discussed many aspects of temperature relating to bioassay determinations. The end observation is always an integration of many individual effects including rates of penetration, intoxification and detoxification, intrinsic insect reaction, etc.

The agriculturally useful new pyrethroids merit a place in this discussion because of their growing and current significance. Temperature control for the bioassay monitoring of pyrethroid residues is an essential. Given below is a table of the response of third instar *Heliothis virescens* larvae to topical application of permethrin (acetone solution) at a variety of temperatures. There are *two orders of magnitude* difference in the LD_{50} in the response *of this insect* to permethrin *applied in the above-described manner.*

The italics above are emphasis that precision is required in the bioassay that includes instar and stage of insect, method of application, nutritional regimen, time of observation, etc. While such very large negative temperature coefficients may be rare, both positive and negative temperature correlations are known for phosphates, carbamates, chlorinated hydrocarbons, etc., in the insecticide field (MATSAMURA 1975, ETO 1974, MENN *et al.* 1957). Similar considerations must be borne in mind for assays based on seed germination or for fungal bioassays and in reality for all chemical-organism interactions. Thus, monitoring for a pyrethroid residue will fail unless there is adherence to strict temperature control for the test (see Table I).

Table I. *Temperature dependence of activity of permethrin when topically applied to III instar, Heliothis virescens.*[a]

Temperature (°C)	LD_{50}	LD_{95}	Slope	No. animals treated
15	0.0017	0.0076	2.6	280
20	0.0063	0.023	2.9	300
25	0.0180	0.059	3.2	260
31	0.0530	0.120	4.6	220
37	0.1100	0.350	3.1	180

[a] *Zoecon* internal data (1979): G. STAAL *et al.*, investigators.

At times we have observed considerable differences in the bioassay due to variables such as relative humidity, light spectrum and intensity, atmospheric pressure, pH of media, population density, age and state of the plant used in the test, formulation compositions, other chemical treatments, etc. These are specified in individual test protocols and in some of the aforementioned general reviews (BUSVINE 1971). For monitoring, where a broad variation of response is satisfactory, awareness of these effects is required for the use of bioassay for precise quantitative determination of residue, strict control of all these variables is absolutely essential.

The above remarks are pertinent particularly for those assays based upon a previously constructed standard curve. When the standards and the unknown are performed simultaneously and under precisely the same conditions variations of temperature and other environmental factors become much less significant.

d) Intoxification and detoxification

It is well known that many pesticides (insecticides, fungicides, and herbicides) behave biochemically as precursors to other substances with biological activities that range from nil to very high. Intoxification is responsible for the high efficacy of the insecticides parathion and malathion (METCALF and MARCH 1953, METCALF 1955, MENN 1978). The systemic fungicide benomyl loses a butyl carboxy group to provide the carbendazim at least equally fungitoxic (CORBETT 1974), and the thiolcarbamate herbicides (Eptam, etc.) are transformed to active oxidized species (CASIDA et al. 1974).

Bioassay methods have led directly to the discovery of toxic metabolites. The bioassay of aldrin-treated animals gave dramatically higher bioassay residues in the fat of rat, sheep, pig, and beef than residues determined by a colorimetric aldrin procedure. Dieldrin does not respond to this colorimetric test. The animal fats contained a toxic metabolite, later identified as dieldrin (BANN et al. 1956).

Whereas chemical assays can usually identify both the parent compound and, if need be, all known metabolites, the bioassay provides an integrated response which is organism- and condition-dependent. For certain objectives, monitoring by bioassay provides a shortcut to a useful number. Considerable background knowledge, however, is necessary for the complete interpretation of that number. Sophisticated assistance in the precise choice of assay method, organisms, experimental conditions, etc., for field monitoring purposes is a necessity. With that assistance, the specific test can be performed in the field to answer one or more well-defined questions such as: Is there toxic hazard to reentry to the field at a specific time interval from spray by the farmer? Or, does the residue in or on the raw agricultural commodity pass or fail the regulatory requirement, or is there reasonable doubt concerning that residue as measured by the monitoring assay?

Note the caveat that these questions and their answers depend both on scientific sophistication in the setting-up of the protocol and the precise execution of that protocol in the field. Unfortunately, many simple bioassays that may be currently applied do not meet these two requirements.

e) Natural bioactive substances

Many plants contain insecticidal and/or insect repelling or other bioactive substances. In fact, many useful pesticides have been derived therefrom: nicotinoids, pyrethroids, rotenone, etc. In soil there are natural fungicidal and fungistatic substances and in reverse there are fungi that severely interfere with the germination and growth of plants.

Care must be taken then in the choice of substrate for the monitoring of residues. LICHTENSTEIN et al. found insecticidal and synergistic substances in dill plants (LICHTENSTEIN 1974b), parsnips (LICHTENSTEIN and CASIDA 1963), turnips (LICHTENSTEIN et al. 1962), and other cruciferous (LICHTENSTEIN et al. 1963) crops. Assays on these crops using flies, mosquitoes, aphids, or other available insects will yield variable residue for organochlorine compounds, pyrethroids, and certain organophosphates and carbamates. These assays will also vary with the maturity of the crop.

Residues depending upon spore germination and mycelial growth of fungi from soil dilutions can give highly unexpected results. In fact, the discovery of many of the antibiotic bactericides and fungicides depends upon the very phenomenon responsible for such unexpected results.

f) Interaction between pesticides

Because bioassay in a monitoring program measures the biological response of an organism or organisms to a summation of all available toxic substances for any given response, the history of the crop or soil treatment and knowledge of specific interactions between the chemicals and the organism is essential.

For repeated treatments of the same or different insecticides, the bioassay provides an availability total response number for all of the insecticides and their metabolites. Sometimes various complex interactions exist between disparate substances as the insecticides parathion and DDT, and the herbicidal triazenes, propanil and 2,4-D (LICHTENSTEIN et al. 1973, LIANG and LICHTENSTEIN 1974) with a variety of insecticidal materials.

These soil-plant interactions are time-dependent (binding, metabolism) and can lead to unreal high estimations of the insecticides in soil and water. Diazinon, Dyfonate, phorate, carbofuran, dieldrin, and others also were synergized by the triazene herbicides.

The residue from the rice herbicide propanil and its phytotoxicity increase in the presence of many insecticides (MATSUNAKA 1968). Such interactions will affect bioassays and their interpretation when the test organism is a plant or an insect and emphasize the need for thorough background knowledge preceding the planning and execution of a bioassay. Supplemental analyses, chemical or biological, are necessary adjuncts for sound monitoring.

Discovered some years ago but investigated intensely and developed quite recently in large scale are a group of substances variously called herbicide safeners or antidotes (PALLOS and CASIDA 1978). These substances, sometimes innocuous in themselves, alter the response of a plant species to a given herbicide. There are currently known antidotes to the thiolcarbamates, chloracetanilides, and others. There is intense research effort directed toward the safening of other herbicides. Bioassays using plants will be affected by these compounds.

III. Particular residue monitoring considerations

a) Bioassay of industrial effluents

Taiwan has made enormous strides in recent years in industrialization in general, as well as in the manufacture of chemicals including agricultural chemicals. Both the developed and the developing nations face increasing ecological and toxicological concerns. The bioassay of industrial effluents provides a unique opportunity for the estimation of the hazards involved with industrial production. In fact, bioassay possesses many advantages over chemical analysis. The American Society of Testing Materials (ASTM) recognizes these advantages; whereas previous testing was chemical and physical, we now cite four recent publications devoted to both chemical and *biological* assay of water pollution (DICKSON *et al.* 1977; see also ASTM 1975 a and b).

In most manufacturing plants, it is uneconomic to segregate all the effluents from each individual unit. The total effluent is frequently a complex mixture of species and of the relative quantities of these species.

If tolerance to effluents for aquatic species is an objective for rivers, inland waterways, bays, and ocean areas, their direct bioassay is superior in many ways to the analysis for all of the constituents (many unknown and unsuspected) in the effluent.

As an example, a series of complex treatments, chemical, physical, and biological of the effluent of a very large industrial plant is monitored regularly by testing the viability of sensitive species of fish after the last stage of treatment and before passing into San Francisco Bay (CHEVRON undated).

Methods of this sort provide startling information sometimes about materials previously conceived to be innocuous, such as upper concentration limits of sulfates and phosphates, and ultimately lead to safe methods for effluent treatment.

The varieties of aquatic organisms used in Taiwan, for example, ought to reflect, of course, the living species (type of fish, crustacean, mollusc, etc., normally present in the environment). Many large industrial manufacturing plants currently employ biologists for the purpose of improving the monitoring system, and such monitoring becomes a multidisciplinary exercise. GLASS (1973) discussed bioassay techniques for environmental monitoring including water that involve fish, zooplankton, phytoplankton, algae, diatoms, cell cultures, etc.

b) Solid waste disposal

The same principle but different methodologies are required for solid waste disposal. Plants, insects, and micro-organisms can serve in the monitoring systems. Downstream liquid effluents are monitored as for any liquid effluents. The use of soil columns to simulate runoff is a sometime expedient. The effects of vapors on caged insects above the waste systems provides information relative to the evolving gases.

c) Pesticide manufacturing plant effluents

The manufacture of industrial and agricultural chemicals in particular can be accompanied by another effluent problem. Many pollutants induce visible plant effects in very minute concentrations. Vapors of

2,4-D esters, suspensions, or aerosols of these and other herbicides frequently derived from manufacturing and formulation operations are always a hazard and require constant monitoring.

By placing sensitive plant species, both broad leaf and grasses, at the boundaries and at strategic areas within those boundaries one can obtain useful information as to the source of polluting effluents and the remedies required to eliminate them. For insecticide manufacture caged insects provide similar information. In part, the used of caged canaries and other birds to monitor noxious gases in coal mines is a bioassay that preceded by centuries the discovery of the gas chromatograph.

Such methods do not differentiate between gaseous or particulate (aerosols, suspensions, etc.) contamination but positive results can lead to further experimental design for such differentiation (air filters, bubbling through high absorbing solvents, etc.).

IV. General methods for insecticide analysis

In this discussion, the plant, soil, water, air, or animal tissue has been treated with appropriate solvents, comminuted and extracted if necessary, concentrated, selected, separated from extraneous solvent solubles in the substrate, and presented to the insect in various forms from the "cleaned-up" solution. The intricacy of the cleanup depends on the chemical and the substrate.

Much of the content of the standard references on residue analysis (Gunther and Blinn 1955), the series Zweig (1963 et seq.), and the book series (Gunther and Gunther 1962 to date) Residue Reviews are concerned with the work-up of substrates prior to the specific analytical procedure.

Although in certain cases the monitoring bioassay may not require this manipulation, the questions to be answered by the bioassay and the advice of an experienced residue analyst and a biologist will be required to define the extent of the separations and the cleanup. Dewey (1958), Sun (1963), and Busvine (1971) all discuss procedures involving both lengthy and almost negligible preliminary manipulation.

Once having obtained an extract the insect of choice may be treated topically, or by immersion, or the insect's food may be treated, or the insect may be subjected to the vapours of the toxicant, etc., or through a combination of two or more of these. We will deal selectively with some of the more common and useful procedures.

a) The surface residue approach

An extract in a volatile solvent (ether, methylene chloride, etc.) can be applied from a calibrated syringe to a vial or to any glass and preferably nonabsorbing surface. The solvent is evaporated with a gentle air stream and frequently Drosophilla melanogaster, or Musca domestica, are introduced and maintained for specified residence times. Comparison is made to tubes containing similar films made with known quantities of the toxicant. This general method was developed by LAUG (1946 and 1948), modified by HOSKINS and MESSENGER (1950), and adapted by hosts of investigators thereafter.

Obviously, care must be taken to deposit films equivalently in the standards as with the unknown. The validity of the assay depends on this equivalence. Volatile toxicants can be lost in the extractions and in the film forming. Large cages are to be avoided, for repellency can keep the insects from direct contact. This is a very general method and not restricted to the above diptera and has been used with a wide variety of insects. Very large amounts of insect extractants will generally provide a low result. This can in part be compensated for by including a similar quantity of plant extractives in a control sample into which various quantities of pesticide have been added.

It should be emphasized here that although we are analyzing for insecticides, insects may not be the target organism and frequently such diverse species as fish, or daphnids, or microorganisms, serve well in the bioassay.

All of the above matter in this earlier and well-developed area of bioassay are discussed in depth with numerous references by BUSVINE (1971), DEWEY (1958), and SUN (1957 and 1963).

b) Direct topical

Where the extract can be reduced to a small volume, a measured quantity (micropipette) can be applied topically to lepidopteran larvae, or to flies, cockroaches, etc. This is a simple technique but it requires careful and sophisticated manipulation. Quantitation depends on management of controls, feeding, and environmental conditions, etc.

c) Immersion and mosquito larva techniques

Various land-living organisms either individually, or on plants, or on artificial surfaces may be immersed in solutions of toxicants. BUSVINE (1971) discussed assays for mites, stored grain pests, eggs of bollworms, ticks, body lice, mealybug, and many others, all dipped or immersed in solutions of the toxicant for purposes of bioassay.

Because of the significance to public health and their ubiquity as well as their sensitivity, mosquitoes whose larvae are aqueous inhabitants have become a favorite organism for bioassay of insecticides. Some aspects of these assays follow.

Mosquito larvae are particularly sensitive to small quantities of certain insecticides. Here, the extract is preferably taken up in a water-soluble solvent and added through a calibrated pipette to the jar containing any of the above (or other) water-inhabiting species.

The larvae or the mosquito *Culex, Aedes*, etc. are best added at the same age and instar. If a mixed population is used, care must be exercised to have a similar mix for the control samples. NOLAND and WIL-

COXON (1950) as well as TERRIERRE and INGALSBE (1953) used variations of this method and mosquito larvae have been a favorite organism
for bioassay of residues for many decades including today.

Solving environmental problems utilizes similar methodology but em-
phasizes other target organisms (ANDERSON 1945, KOCHER et al. 1953,
WASSERBURGER 1952) such as daphnids. PAGAN and HAGEMAN (1950)
employed guppies. Frequently, mosquito, fish, trout, crustaceans, and
molluscs are employed in variations of this procedure.

With the current and increasing concerns with environmental problems, governmental regulating agencies have issued guidelines that are
in essence extensions of the above methodologies. For fish, aquatic invertebrates, estuarine and marine organisms, etc., the *Federal Register*
(1978) provides the organisms, best conditions, standards, etc. for toxicant assays. Based upon these standards, private testing companies have
been organized to carry out (at a fee) such testing. Some of them regard
their protocols as proprietary. The bibliography lists one company that
employs daphnids and midges, and another daphnids and fish. These
protocols can be adapted for monitoring bioassays (*SRI International:*
Menlo Park, California; *Aquatic Environmental Sciences:* Tarrytown,
New York).

d) Techniques based on food ingestion

As compared to topical application, food ingestion methods are generally somewhat more complex. Also they generally but not inevitably
require a larger quantity of the toxicant and a longer observation period
for a measurable response. From our internal data at *Zoecon* the following are the comparable doses for Pydrin, a new pyrethroid for topical
vs. feeding methods (see Table II).

Table II. *Pydrin bioassay—topical versus feedthrough.*

Assay	LD_{50}	Days to observation
Topical (3rd instar)	0.01 μg	3
Feedthrough[a] (1st through 4th instar)	0.940 ppm in food	7

[a] Food preparation, maintenance, etc. much more complex for feedthrough assay
(*Zoecon* internal data).

The preparation of the diet must be strictly monitored both for the
control and the treated sample. It is important in a short-term test to
avoid any choice of food source. Thus, if a volatile solvent extract is
used to treat sugar or milk with toxicants, no other source of food materials should be present. SUN and SUN (1953) fed homogenized milk
with toxicant directly to houseflies.

Zoecon regularly employs a feedthru technique with a special artificial diet and with a confined cell for the insects *Heliothis virescens, Manduca sexta,* and others. Direct feeding of homogenized plant materials has been employed by SUN and PANKASKIE (1954). Variations of this technique can also be used on *Tribolium confusum* and other stored grain pests with or without maceration. Such techniques give a response that does not differentiate contact from stomach poison effects, but as long as the positive controls are treated equivalently, the method gives a reasonable approximation of the residue.

For the newer insecticides that are inhibitors of chitin, biosynthesis feedthru techniques are preferable. In fact, in many insects the response to topical application or other contact methods only occurs at very high doses so that bioassay for Dimilin by these techniques becomes difficult. (Parenthetically, the discovery of this new and important group of insecticides required a more sophisticated technique in the routine screening than was generally employed.) *Spodoptera littoralis,* however, is an exception to this general response to Dimilin, but its laboratory culture for bioassay purposes presents a problem.

e) Comparison bioassay and chemical assay

SUN (1957) provided a comparison from many investigators of the bioassay results of pesticides as compared with other analyses either chemical or other types of bioassay. The plants in this comparison include dock, marrow, cabbage, spinach, peaches, celery, onion bulbs, potato foliage and tubers, corn (including foliage), stringbeans, yams, turnip, and carrots, as well as various soil samples, butter fat, rat fat, and beef fat. The pesticides in this comparison include DDT, aldrin, dieldrin, BHC and lindane, chlordane, toxaphene, heptachlor, parathion, and malathion. In only an exceptional case the ppm found differ by 100%. Most of the results are reasonably close and equivalent to the variations an experienced residue analyst finds in his normal assay work. It should be noted that none of the toxicants were highly systemic and that the removal of the toxicant from the substrate (soil or plant or animal tissue) was the same in each case. Further, the residues were quite high as compared with the standards permitted today. In another example a large number of extracts were fortified with the equivalent of 200 ppm of dieldrin (unpublished *Shell* data quoted by SUN 1957) and gave equivalent results in a dry film assay (housefly, Drosophila) and a spray method (larvae, *Aedes aegypti*). (Had aldrin been the fortification agent and a finite residence time permitted, the results would not have been the same as has been indicated earlier.)

f) Bioassay of insecticidal soil residues

Topical, immersion, film and other methods have been applied to soil residues. EDWARDS *et al.* (1957) employed *Drosophila melanogaster* for the assay of aldrin and lindane in midwest soils. In a study of the effects on a variety of crops of high soil content of aldrin and heptachlor, LICHTENSTEIN (1960 b) analyzed both soil and crop residues (including the epoxide) by bioassay using the fruit fly. The interaction of herbicides and insecticides (triazenes, 2,4-D, etc. with parathion, DDT, etc.) in soils was measured by LICHTENSTEIN et al. (1973) and LIANG and LICHTENSTEIN (1974). This study provides a warning in regard to need for knowledge of the agricultural treatment prior to the planning or interpretation of the assay. The bioassay of crops grown in aldrin and heptachlor-treated soils was determined by use of *Drosophila* by LICHTENSTEIN et al. (1965). The translocation of 6 organochlorine compounds from treated soil into the aerial shoots of pea plants was similarly studied by LICHTENSTEIN and SCHULZ (1960).

The ecological problem of the persistence of organochlorine toxicants in soil was studied by a direct feeding method (no extractions) or by a dry-film procedure (LICHTENSTEIN et al. 1960). Frequently used for nematicide and stored grain pests, exposure to vapours can produce insect toxicity. Soil columns are part of the methodology of the experimental agronomist. Employing a combination of volatility and a soil column, HARRIS and LICHTENSTEIN (1961) bioassayed various insecticides and such soil variables as moisture content, air flow, etc.

Methylenedioxyphenyl synergists added to the soil inhibited microorganism conversion of aldrin to dieldrin (LICHTENSTEIN et al. 1963) as measured by both bio- and chemical assay. Bioassay of various forms of azinophosmethyl provided information on both persistence and metabolic products (SCHULZ et al. 1970). These studies also illustrate the use of bioassay in combination with TLC (or GC or HPLC) for the identification and quantitative estimation of bioactive fractions. In 1950 this author utilized bioassay similarly in the process development for the manufacture of TEPP (unpublished *Chevron* chemical files). A bioassay method provided the information on the translocation and metabolism of phorate in a model ecosystem (LICHTENSTEIN et al. 1974 b). In fact, the METCALF (1972) and other models of ecosystems can be utilized for quantitative bioassays.

g) *Attractants and repellents*

All bioassays for insecticides are complicated by particularities of the insects' behavior. The actual bioassay for attractancy and repellency requires very complex procedures and is beyond the scope of this review.

Bioassay is, however, the principle technique employed in the identification and discovery of attractants (INSCOE and BEROZA 1976). Recognition of a repellency or attractancy factor is important for those insecticide-monitoring methods based upon nontopical application methods. It is well known that results based upon food ingestion or contact can be seriously biased by repellency factors. For the pyrethroids and for many other insecticides this repellency has been widely documented. Good bioassay design, as for example the small vial employed by HOSKINS and MESSENGER (1950), reduces errors caused by repellency. BUSVINE (1971) devotes a chapter on the treating of repellents and attractants; so far, the need for monitoring their residues is slight.

h) Stored grain testing (flour bins, containers)

Huge losses are incurred after harvest for many crops. Rodents, insects, fungi, yeasts, and bacteria attack stored grain, peanuts or ground nuts, fruit, etc. These are frequently treated by standard biocides before or after harvest. In certain countries seed so treated must be easily recognized and differentiated from the seed used for food.

The question frequently arises then: Does this stored product or its container or the seed have pesticide residues? BUSVINE (1971) discussed many of the older bioassay methods for such determinations. A useful assay procedure is found in the fungicide section of this review.

In general, a susceptible insect, e.g., *Tribolium confusum*, is introduced into the acceptable stored substrate. After a given time interval the number of surviving insects is measured and compared with a non-treated control. The same general test can be made of treated burlap or cardboard or wood containers. In an alternate procedure the stored product is extracted and the extract is tested in any of the procedures discussed under insect testing. By use with a very sensitive species small residues may be relatively rapidly assayed.

i) Bioassay for insect growth regulators (IGRs)

Currently, cisterns, stagnant water pools, sewers, etc. are treated with insect growth regulators (IGRs) usually from slow-releasing formulations. To answer the question of what quantity of IGR is present at a given time, third instar larvae of *Culex* or other available species are added to the water sample (or to an extract of a substrate to be analyzed) and the development through pupal and adult form observed. This is compared with a similar observation of various concentrations of the IGR substance to be assayed. With the usual reservations, a residue estimate can be made. SCHAEFFER and WILDER (1971) provide a practical bioassay for both IGR and standard insecticides.

Another example of a useful bioassay of IGRs relates to the treatment of cattle through their food intake for the control chiefly of diptera, such as the hornfly and the stablefly. The IGR passes through the alimentary canal and a residue is found in the manure. In this case, larvae of the hornfly (a similar procedure can be used for housefly larvae) are introduced into the manure and their development noted. The bioassay closely correlates with an actual chemical analysis and provides an efficiency test for formulation of methoprene currently used for hornfly control in the United States (HARRIS et al. 1974).

V. Bioassay of fungicidal and bactericidal residues

Although in much temperate zone, dry area agriculture, as in the valleys of California in particular, foliar fungicides have limited application; in wet and tropical areas plant disease caused by these organisms is a severely limiting productivity factor. All soils that are agriculturally useful are storehouses for a multiplicity of organisms. Although fungicides roughly account for less than 20% (Kohn 1974) of the sale of agricultural pesticides, they are currently the most rapidly expanding of the three main classes mostly because of needs in the developing countries.

Significantly, in the United States almost 100% of corn seed is treated with fungicides to control *Pythium ultimium* and other pathogens. Large proportions of the seed of cotton, small grain, rice, and vegetable crops are similarly seed or soil-treated with fungicides and other toxicants and the monitoring of such residues is part of the content of this section of the review.

Horsefall (1945 and 1956) and Torgeson (1969) discussed in varying detail aspects of organism-chemical interactions that played a role both in the discovery of the toxicants and as an indication of the presence of the toxicants (basis for assay procedures). The discovery of penicillin by chance (fungal contamination of an agar plate supporting a bacterial population) has become popular folklore and continues to be the basis of one of the most widely used assay methods (inhibition of mycelial growth).

a) The spore germination test

In general, a solution or suspension is prepared by the extraction of the substrate to be assayed. The soil, plant tissue, water, or air is extracted by appropriate solvents (Torgeson 1969 *et seq.*, Kelman 1967). The chemical is deposited upon a demarcated area of the slide (wax pencil, glass cutting stylus, etc.). Special slides with a pre-fabricated depressed area can be purchased. The solvent is removed, the toxicant remains. A droplet containing a spore suspension is added and observations are made with the aid of optical magnification of the germination, or lack of, of the particular spores.

The spore germination method has been a standard procedure in mycological laboratories for many decades. Much attention has been given to statistical analysis and improvements of the methods. Test-tube dilution methods (with extract) provide easy approaches to a standard curve. Frick (1964) specifically discussed methods of reducing variability in the results of glass slide spore germination assays of fungitoxicity and that discussion is useful for the bioassayist. This is a sensitive assay procedure, simple in its general outline but requiring careful controls of the organism source, culturing conditions, temperature, and other variables (Neeley 1969).

b) Mycelium inhibition procedures

In general, a culture of a given fungal or bacterial species is grown on a medium and under environmental conditions optimum for that organism. Extracts of crop in soil are added to the culture medium by a variety of techniques and observations are made generally over periods from 1 to 7 days.

The above type of test can be adapted for most easily cultured organisms. Discs of filter paper or other absorbent may be used for extracts from soil or foliage. The choice of organism for culture depends upon the toxicant to be analyzed. The most sensitive species, if easily cultured, is the optimal choice. The execution of the analysis of the unknown, simultaneous with the controls and standards, eliminates much of the error resulting from variations due to cultural and environmental differences. The advice and availability of a professional plant pathologist for the setting up of these tests, the source of innoculum, and interpretation of results of initial assays guarantees increased reliability from this procedure.

Sometimes the agar is made up with a serial dilution of fungicide or other toxicant and the organism introduced thereto. BATES et al. (1962) investigated the 2-methyl-4-dinitrophenols. Petri dishes with potato dextrose agar are made up containing 0, 4, 20, 100, and 500 ppm of toxicant. Mycelial growth of Pythium ultimium, Sclerotinic fructogenea, Verticullum alboatrum, and Alternaria solani and conidia of Aspergillus niger and Botrytis cinerea are observed. Concentrations providing a reduction of colony diameter after innoculation and growth period by 50% are the basis of quantitation.

Another variation of the mycelium inhibition test was described by BARKAI-GOLAN and LATTAR (1962) in which captan suspensions are employed. Here radial growth measurements are observed for 4 days at 25°C for a large variety of fungi. This methodology is applicable to the testing of all the sulfenimides and many other fungicides. It is especially useful for insoluble or low-soluble and solvent-labile toxicants.

The monitoring of certain fungicidal residues by bioassay methods must take into account, as always, the peculiarities of the physical and chemical properties of the toxicant. KOHN (1977) discussed the lability of the N-S bond in the sulfenimide fungicides captan, Difolatan, Phaltan, etc., particularly when in solution. Solvents, extraneous substrate solutes, and residence time in the solution before assay will decisively affect both spore germination and mycelial inhibition assays.

c) Crop applications

LUKENS (1975) recently determined Cu^{++} and cycloheximide residues on tomato plants using Alternavia solani as the test fungal organism. He used leaf discs from treated plants directly without extraction.

Similarly (LUKENS 1976), used this leaf disc method to determine directly the residues of chlorthalonil. The discs are stored at 25°C and observations are made over a period up to 17 hr.

Both of the above methods may be quantified by comparison with treated substrates to be analyzed and sometimes foliar tissue itself or seeds are then applied to the culture by a variety of methods. References are given in the fundamental texts (TORGESON 1969, LUKENS 1971, HORSEFALL 1957). The American Phytopathological Society has published (KELMAN 1967) simplified methodology that can be the basis for some meaningful assays. For example, MITCHELL (1967) takes treated seed and measures the inhibition zone surrounding the seed pressed into a potato dextrose agar culture of *Glomerella cingulata*. A standard curve is constructed from untreated to graded treatments of the seed through this measure of the radii of the inhibition zone.

For foliar fungicides, MUNNECKE (1967) takes discs of leaves (from various plant species) and places them on a culture of *Myrothecium verrucaria*. In this way Cu^{++} and the dithiocarbamate fungicides can be assayed after a standard curve has been constructed. The culture is maintained at 20°C.

Mention should be made of certain specialized assays. In general, these require greater sophistication and are, hence, perhaps more limited in their usefulness. Soils and plants are sometimes treated for *Phytophora parasitica* and other phytophores. They are normal constituents in semitropical soils (sometimes a limiting factor for pineapple seedlings); a reduction of their population through culture methods may indicate the presence of a soil fungicide. TSAO (1960) developed a serial dilution endpoint method for estimating disease potentials of *Citrus phytophoras* in soil. By comparing untreated soils with treated soils and measuring the brown rot citrus lesions, a rough bioassay may be established. Such testing is complex and requires perhaps too great plant pathological technical competence for transfer into a simple monitoring procedure.

Diphenyl and the chlorinated nitrobenzenes may be assayed by another variation of the mycelial growth method (GEORGOPOULOS and VOMVOYANNI (1965). Serial dilutions of these chemicals are made and added to the warm potato dextrose agar in Petri dishes. Plugs from cultures of *Hypomyces solani* (f. cucurbitae) are placed in Petri dishes. The observations of the growth of the colonies are made at specified intervals by measuring the length of perpendicular radii.

Although pharmaceuticals are considered outside the scope of this paper, bacterial and algal rather than fungal organisms can serve in bioassays of pesticides of all varieties. The growth of certain bacteria in liquid cultures can provide an easily quantified turbidimetric method for quantitation. Here again assistance by bacteriologists in the setting up of the tests is recommended because of the specificities of the population growth characteristics and the need to control the usual physiological and chemical variables. For bacterial methods, SHENNAN and

FLETCHER (1965), TCHAN et al. (1975), and TCHAN and CHIOU (1977) are useful sources of information for tests of fungicidal and nonfungicidal toxicants.

Before concluding this section on microorganism testing, one caveat is appropriate. Contamination may be a source of serendipitous discovery but it can be a plague to the bioassayist.

VI. Herbicide bioassay

Although insecticides preceded herbicides in therapeutic development, it is well known that the latter currently dominate the agricultural market place. Plant science is an active and expanding field. Nowhere is this more evident than in the bioassay for plant-affecting substances. The diversity of assay procedures is extraordinary. The table which follows describes with references laboratory assays for 21 herbicides, many of which are representatives of large groups of economically significant compounds. TRUELOVE (1977) described in partial detail the steps for many of these assays and provided numerous references from which the complete procedure can be obtained (both chemical and biological). The chapters in this volume by SANTLEMAN on bioassay in general, by MORELAND on reactions involving isolated chloroplasts, by TRUELOVE and DAVIS on isolated mitochondria, and others are provocative, current, and provide techniques for bioassay or suggest adaptations for novel and exploratory bioassays. The details for the culturing, maintenance, and quantitation of the algal approach to herbicide bioassay is also provided (see Table III).

Photosynthetic inhibitors which include some of the most widely produced and economically exploited herbicides can all be assayed by a relatively simple technique that easily lends itself to a monitoring program. TRUELOVE et al. (1974) described this potentially useful technique briefly. They cut discs from pumpkin seedling cotyledons and floated them in cups or jars. The floating discs are illuminated under controlled conditions. The discs continue to float as long as gas exchange from photosynthesis continues. In the dark, or if photosynthesis is inhibited, the discs sink. Observations are made for floating or sinking over a 12-hr period. With care and with enough seedlings for 10 discs per given concentration, a standard dose response can be obtained. Prometryne gave a positive response from 0.002 to 0.02 ppm by this method. Atrazine, bromacil, fluometuron, and diuron, all gave useful responses. The test was negative for nonphotosynthetic inhibitor herbicides and, hence, is mechanistically selective. Cucumber cotyledons may be substituted for pumpkin.

DA SILVA et al. (1976) slightly modified this procedure using watermelon (Citrallis vulgaris) cotyledons incubated and illuminated at 27°C for 15 to 24 hr. Paraquat and diquat are less active than the triazines and ureas which provide positive responses at $10^{-7}M$.

116   G. K. KOHN

Table III. *Bioassay species used for selected herbicides.*

Herbicide	Bioassay species and reference[a]
Alachlor	Japanese millet (2), *Chlorella* (102).
Amiben	Soybean (13), cucumber (5,31), oat (75), giant foxtail, velvet leaf (92).
Atrazine	Oat (58,80,88), soybean (13), cucumber (26,80), *Chorella* (40, 102), green foxtail (47), Japanese millet (2), pumpkin (97).
Bromacil	Pumpkin (97), watermelon (25).
CDEC	Corn, cucumber (33), oat (69,86), cotton (99).
Chlorpropham	Cucumber (31,43), barley, mustard (12,84), oat (17,22), crab-grass (84).
Dalapon	Millet (40), barley, mustard (12), oat (21,46), cotton (99), cucumber (20).
Dicamba	Soybean (15), bean (89), oat (37), sorghum, morningglory (74), cucumber (20,29).
Diphenamid	Oat, barley, ryegrass (30,37,42), tomato, pigweed, lambsquarter, rape (34).
Diuron	Cucumber (26,29,63), ryegrass (98), barley (19), *Chlorella* (1), oat (4,46,58,87), duckweed (4), pumpkin (97).
Dinoseb	Corn, soybean (28,59), cotton (99), cucumber (26), watermelon (25).
EPTC	Oat, millet sorghum, ryegrass (24), cucumber (55,63), pumpkin (97).
Fluometuron	Oat, millet sorghum, ryegrass (24), cucumber (55,63), pumpkin (97).
Linuron	Ryegrass, cucumber (30,53), sugarbeet (27), duckweed (73), sorghum (77).
Monuron	Crabgrass (67), oat (87), soybean (103), barley, mustard (12), ryegrass (39), *Chlorella* (1).
Paraquat	Duckweed (32), wheat (94), cucumber (26,101), bean (7), watermelon (25).
Picloram	Bean (38,51), cucumber (20,29,55), sesbania (37), lettuce (81), sunflower (56).
Prometryn	*Chorella* (1), sugarbeet (27), cucumber (55), oat (82,104), wheat (100), pumpkin (97), watermelon (25).
Simazine	Oat (85,88), ryegrass (39), mustard (12), soybean (85), duckweed (70), sugarbeet (27).
Trifluralin	Morningglory, alfalfa, velvetleaf, foxtail (74), sorghum (31,60, 74), oat (18,44,83), barley, cucumber (6,20).
2,4-D	Cucumber (31,76,90), crabgrass (67), cotton (52), mustard (50), duckweed (10), tomato (50,64), sesbania (35), soybean (103), *Chlorella* (40).

[a] See TRUELOVE (1977) for references.

REID and HURTT (1969) described a rapid bioassay for the identification and quantitation of picloram. An older work (READY and GRANT 1947) provided the basis of assay for 2,4-D in aqueous solution. Bioassays based on *Cucumis sativa* and *Sorghum vulgara* (cucumber and sorghum) are particularly applicable to soil and leachates and extracts. The process is described by ESHEL and WARREN (1967). Diverse herbicides such as 2,4-D, Amiben, CIPC, Trifluralin, and others can be monitored by this procedure.

A rapid and extremely sensitive bioassay has just been reported in *Chemical Abstracts* (March 5, 1979). O'BRIEN and PRENDEVILLE (1978) observed the leakage of electrolytes through the membranes of *Lemna minor* (duckweed) after 3 hr and after 72 hr exposure to diquat and paraquat. After 72 hr 0.00018 micrograms/ml of diquat and 0.000017 micrograms/ml of paraquat give a measurable response!

The comparative value of bioassay and instrumental analysis of herbicides has been recently evaluated by EBERLE and GERBER (1976). HOROWITZ (1976) and SAGGERS (1976) relate the search for new herbicide to bioassay techniques. The direction of this research, as previously indicated, is in part away from the earlier whole plant assays and toward more specific special organisms and tissues. ZILKAH (1977) and collaborators examined plant cellular tissue, single cell suspension, and plant tissue culture in general. Toxicity in seedlings and in tissue culture is compared. ASHTON et al. (1973) described protoplasts as an aid to localizing the site of action of herbicides. A very new approach to bioassay that transcends herbicides is the investigation and assay of bioluminescence of bacteria (TCHAN and CHIOU 1977). SHENNAN and FLETCHER (1965) discussed earlier 2,4-D interactions with microorganisms. This isolated and specialized organism assay, because it relates more closely to mode of action, is particularly useful for the development of plant growth regulators.

a) Algae

Algae and certain simple lower aquatic plants have frequently been employed in the screening of herbicide candidates. The conditions for the optimization of the organism's growth and the relative ease attached to the control of those conditions recommend algae for residue-monitoring assays.

Summarized in Table IV are some references relating to groups of pesticides and their algal interactions which have been used in bioassays for pesticide residues (McCANN and CULLIMORE 1979).

Table IV. *The use of algae in bioassay for pesticides.*[a]

Pesticides or types of pesticides	Reference
Atrazine	ATKINS & TCHAN (1967)
Atrazine, monuron, diuron, neburon	TCHAN et al. (1975)
Triazenes, phenylureas, miscellaneous herbicides, N-methyl carbamates (insecticides)	HELLING et al. (1974)
Diuron, etc.	ADDISON & BARDSLEY (1968)
Photosynthetic inhibitors, respiratory inhibitors	KRATKY & WARREN (1971)
Barban, chlorpropham, propham	WRIGHT (1975)
Herbicides (many varieties)	NOLL & BAUER (1973)

[a] From information provided by McCANN and CULLIMORE (1979).

In addition, the above-cited review discusses many more interactions between a broad variety of algal species and herbicides, insecticides, fungicides, and bactericides that could be adapted to bioassay.

Many algal species (*Chlorella, Phormidium, Nitella, Chlorococcum, Hormidium, Palmella, Nostoc,* and others) have been employed in these assays. Their sensitivity and usefulness have been compared to the more familiar oat seedling test and seed germination test. Not infrequently insecticides, fungicides, nematicides, as well as herbicides have measurable effects upon the growth of these algal cultures. This provides both an opportunity and a hazard relating to algal assays. The history of treatment should be known. Complementary or supplementary chemical and biological assays may be required.

For water assays the direct growth of the algal species in the sample to be treated is compared with the growth in the waters with simulated equivalent organisms and nutrient composition. Dose-response ratios over a range significant for the unknown to be assayed give the basis for the quantitative estimation. For waters with dissolved organic suspensions of silicas, aluminas, clays, etc. the bioassay gives the available toxicant by these assays, not the total quantity in the sample.

1. Filamentous algae and small aquatic plants.—In this respect, *Nitella* and *Elodea* (a plant) may be used for diquat estimation (CHEVRON, internal publication) for pond and river waters. Care is taken to provide a standard nutrient regimen. This assay can utilize easily obtainable materials that can substitute for sophisticated chemical laboratory apparatus. Aquatic plants such as *Lemna minor* (duckweed) can substitute for (or supplement) *Elodea* as can other easy-to-culture algal species. Because of this adaptability and simplicity of execution, aspects of this assay will be briefly discussed.

α. Residue application.—Run-off waters, stream, or lake water can be used directly. The nutrient supplement should be added to provide an approximately equivalent growth potential.

Residues from crops may be extracted. A final acetone extract is preferable for these procedures. Any of the usual crop extraction procedures, as from standard texts previously cited, may be used. To avoid additional turbidity, the solvents in these procedures, if not water-soluble, should be removed and the residue taken up in acetone.

Similarly, for soil residues extractions may be performed. Soil columns and elutions therefrom may be used directly depending upon the nature of the toxicant to be examined. If digestion procedures and extractions are employed, one should consider that metallic ions (Cu^{++}, for example) are extremely toxic to algae and, as will be noted, Cu^{++} can be used as a positive control.

β. Further aspects of an algal or small aquatic plant assay for herbicidal compounds.—Cultures of the *Elodea* or similar species are maintained in 1- to 5-gal jars. To maintain these cultures, a nutrient solution is added. Several days prior to use, 4- to 6-cm sections are placed in

1- or 2-gal jars in a sunlit area containing ¼ strength Hoagland's Solution or similar nutrient mix.

Similarly, for *Lemna minor*, 1 to 50 individual fronds can be used instead of, or to supplement, the algal species. After the above preparation 1 section of *Elodea* about 5 cm in length, or 20 fronds of *Lemna*, are placed in 150 ml of water in a polystyrene cup containing the nutrients. Depending upon the algal variety and the temperature, this test requires a 1-day to 1-wk growth period after the introduction of the residue solution.

γ. *Observations.*—After the extract is added, observations are made of various indications such as rate of growth, absolute increase or decrease of size at a given time, and, of course, life or death of the organism.

These observations can be simplified by working in the known lethal ranges of the suspected toxicants. This assay, as for most bioassays, is excellent as an indicator for gross contamination caused by algicidal or aquatic herbicides. It does not identify individual chemical species or tightly bound pesticidal species.

The tremendous scope of algal assay methods for pesticides can be concluded by examination of the tables abstracted from McCann and Cullimore (1979) (See Tables V and VI).

Note that though herbicides predominate, certain insecticides and fungicides also interact. In the opinion and experience of this reviewer the algal and aquatic plant assay procedures are simple to perform, sensitive, and will be more utilized in the future.

The *in vivo* methods reflect upon the populations in soil and water

Table V. *Summary table of pesticide toxicity to algae* (I).

Toxicity level	*In vivo* tests
Highly toxic to at least some algae (<0.1 ppm)	Diuron
	Monuron
Toxic to some algae at field application levels (approx. 0.1 to 5 ppm)	Atrazine
	Linuron
	Mancozeb[a]
	Metribuzin
	Pentachlorophenol
	Prometryne
	Sodium pentachlorophenate
	TCA
Toxic to most or all algae at high concentration (>5 ppm)	Simazine
	A-18-17
	2,4-D
No toxicity demonstrated	Methabenzthiazuron
	Picloram
	Siduron
	TCA
	2,4,5-T

[a] Fungicide or insecticide.

Table VI. *Summary table of pesticide toxicity to algae* (II).

Toxicity level	In vitro tests		
Highly toxic to at least some algae (<0.1 ppm)	Barban[b] Bromacil Diuron EPTC[b]	Fenuron Fluometuron Linuron Metobromuron[b]	Metribuzin Monuron Neburon
Toxic to some algae at field application levels (approx. 0.1 to 5 ppm)	Amitrol-T[b] Asulam Atratone[b] Atrazine[b] BHC Barban[b] Chlorpropham Dalapon[b]	Dicamba[b] Dichlobanil[b] Dinoseb Dinquat MCPA[b] Paraquat[b] Pentachlorophenol Picloram	Propanil Propazine Sodium penta- chlorophenate Simazine TCA 2,4-D[b] Trifluralin
Toxic to most or all algae at high concentration (>5 ppm)	Amitrol-T[b] Atrazine[b] BHC[a] Bromoxynil Ceresan[a] DDT[a]	Diquat[b] Ioxynil M & B 8882 Metobromuron[b] Methoxychlor[a]	Mirex[b] Paraquat[b] Siduron Simazine[b] 2,4,5-T
No toxicity demonstrated	Barban[b] Chloronil[b] Dalapan[b] Dicamba[b] Dichlobenil[b]	Diphenamid EPTC[b] MCPA[b] Malathion[a]	Solan 2,4-D[b] 2,4-DB Vapam[a]

[a] Fungicide or insecticide.
[b] Different studies have shown widely divergent toxicity levels for the pesticide listed.

as affected by the toxicants while the *in vitro* methods provide the basis of most bioassays. This list, although extensive, does not include interactions with many fungicides, bactericides, or antibiotics because some of these effects remain unpublished. Very recently a report on the effect of 2,4-D and 2,4,5-T has appeared (Cox and BOARDMAN 1978). *Microcystic aerninosa*, a blue-green algae, is stimulated by these herbicides at concentrations down to $10^{-8}M$. At $10^{-3}M$ the herbicides became phytotoxic. Such observations have been made with fungicides and bactericides and zenobiotics, and even with plant phytotoxicity to H_2S and SO_2 (THOMPSON and KATS 1978). Where such nonlinear and discontinuous functions exist the bioassayist must carefully regulate his choice of concentrations.

Quantitation in most algal studies is carried out with turbidimetric measurements (liquid cultures), chlorophyl extraction, or cell counts.

b) Plant growth regulators (PGRs)

It is common knowledge that, by altering the rate and mode of application of many herbicides, useful plant growth regulating properties may be obtained. Small quantities appropriately applied of the phenoxy acids can be used for fruit tree thinning. Similarly, glyphosate or the diquaternaries can affect cane flowering and sugar production.

Much current research effort is devoted to growth regulants (PGRs) although their impact on current agriculture is small. With an eye to the future rather than to the present monitoring needs, this section on plant growth regulants is written.

WEAVER (1972) and AUDUS (1965), and the PGR Working Group (1977) all presented general treatments that include bioassay of PGRs. MITCHELL and LIVINGSTON (1968) provided specific assay methodology. GLASS (1973) in his discussion of environmental impact provided monitoring references for PGRs as did the EPA (1976) and GAMBORG (1975). Many of the methodologies including algae, bacteria, and whole plants can be judiciously applied for monitoring of this interesting group of chemotherapeutants, the plant growth regulators.

Summary

The physical and chemical estimation of pesticide residues is scientifically advanced technology. The ability to detect minute quantities of toxicants in complex substrates has had tremendous impact upon science, medicine, government, and politics. However, the methodologies for these analyses generally require the use of highly sophisticated and quite expensive equipment.

The bioassay has been an invaluable tool for the discovery of new compositions and is an aid in the elucidation of mechanism. Highly sensitive bioassays are now available. The biological assay frequently provides useful although different information than that derived from the chemical analysis.

The implications of this useful information have been explored. The limitations and advantages have been described. The chemical and the bioassay can be complementary and supplementary. The procedures for obtaining the extracts to be analyzed (the work-up) are frequently the same.

This highly selective review has examined various procedures and the constraints, limitations, advantages, and disadvantages of the bioassay. We have chosen examples of useful assays from both the older and the most modern procedures. Because it is our conviction that bioassays will continue to expand in their usefulness, the criteria for the practical extension of bioassays are enumerated:

(1) A good bioassay requires a multidisciplinary approach which should include at least one experienced residue chemist and one trained biologist.

(2) Whatever the class of toxicant to be analyzed, the biologist should consider organisms from all the phyla for his assay. Although plant reactions are a normal choice for bioassay of a herbicide, a microorganism may provide a simpler, more sensitive, and rapid procedure.

Similarly, a fish or crustacean may be as useful in the bioassay of an insecticide as would the normal insect test.

(3) To reduce the effects, extrinsic and intrinsic (environmental and organism variables), the unknown and the standards should be tested with as nearly identical procedures and as simultaneously as is possible.

(4) The interpretation of bioassay results to be meaningful requires both the biologist and the chemist. The test must be validated from time to time by either another bioassay or a chemical analysis.

(5) One should design a bioassay to answer a specific question or group of questions. For example, "are there toxic contaminants in a given water supply?" This may be answered by a simple fish or crustacean assay. The source of this contamination can be a potent toxicant or merely a higher than normal concentration of inorganic ions. Other assays are required for more detailed and particularly more specific information.

(6) In general, assays become more meaningful and simpler to perform if indigenous organisms are substituted for the cited organisms.

(7) Regulatory complexities will increase the future usefulness of bioassays.

The principles for all of the above have been emphasized and discussed in varying degree in the body of this paper. Finally, this review has selectively explored certain bioassay and monitoring procedures for insecticides including insect growth regulators, fungicides and bactericides, and herbicides and plant growth regulators. Organisms employed are as varied as bacteria, fungi, algae, many varieties of insects and mites, many kinds of plants in various stages of their development, certain fish, and crustaceans with brief reference both to other types of agents and other species. The methodologies embrace substrates as diverse as soil, industrial wastes, animal wastes, living tissue from plants, or from animals, air, and water. Both research objectives and monitoring needs will require increasing future use of bioassay.

References

Addison, D. A., and C. E. Bardsley: *Chorella vulgaris* assay of soil herbicides. Weed Sci. 16, 427 (1968).
Anderson, B. G.: Toxicity of DDT to daphnia. Science 102, 539 (1945).
Ashton, F. M., and A. S. Crofts: Mode of action of herbicides. New York: Wiley-Interscience (1973).
ASTM Publication: Water quality parameters. Philadelphia: ASTM STP 573 (1975 a).
—— Water pollution assessment—Automatic sampling and measurement. Philadelphia: ASTM STP 582 (1975 b).
—— Biological methods in the assessment of water quality. Philadelphia: ASTM STP 528 (1973).
Atkins, C. A., and Y. T. Chan: Study of soil algae. VI. Bioassay of atrazine and the prediction of its toxicity in soils using an algae growth method. Plant and Soil 3, 432 (1967).

Table VII. *Chemical designations of compounds mentioned in text.*

Name used in text	Chemical designation
Alachlor	2-Chloro-2'-6'-diethyl-N-(methoxymethyl)-acetanilide
Aldrin	Hexachlorohexahydro-*endo, exo*-dimethano-naphthalene
Amiben	3-Amino-2,5-dichlorobenzoic acid
Amitrole	3-Amino-1,2,4-triazole
Asulam (Asulox)	Methyl sulfanilylcarbamate
Atrazine	2-Chloro-4-ethylamino-6-isoproylamino-S-triazine
Azinophos-methyl (Guthion)	O,O-Dimethyl S-{(4-oxo-1,2,3-benzo-triazin-3(4H)-yl)methyl} phosphorodithioate
Barban	4-Chloro-2-butynyl N-(3-chlorophenyl) carbamate
Benomyl	Methyl 1-(butylcarbamoul)-2-benzimidazole carbamate
BHC	Benzenehexachloride
Bromacil	5-Bromo-6-methyl-3-S-butyl uracil
Bromoxynil	3,5-Dibromo-4-hydroxy-benzonitrile
Captan	*cis*-N-[Trichloromethyl)thio]-4-cyclohexene-1,2-dicarboximide
Carbendazim	2-(Methoxycarbonylamino)-benzimidazole
Carbofuran	2,3-Dihydro-2,2-dimethyl-7-benzofuranyl methyl carbamate
CDEC (Vegadex)	2-Chloroallyl-diethyldithiocarbamate
Ceresan	N-(Ethylmercury)-p-toluenesulphanilide
Chlordane	1,2,4,5,6,7,8,8-Octachlor-2,3,3a,4,7,7a-hexahydro-4,7-methanoindane
Chlorpropham	Isopropyl N(3-chlorophenyl) carbamate
CIPC (Chloro-IPC)	Isopropyl m-chlorocarbanilate, or isopropyl-N-m-chlorophenyl-carbamate
Cycloheximide (Acti Acid)	3-{2(3,5-Dimethyl-2-oxocyclohexil)-2-hydroxyethyl}-glutarimide
2,4-D	2,4-Dichlorophenoxyacetic acid
Dalapon	2,2-Dichloropropionic acid
DDT	Dichloro diphenyl trichloroethane
Diazinon	O,O-Diethyl O-(2-isopropyl-6-methyl-4-pyrimidinyl) phosphorothioate
Dicamba	3,6-Dichloro-2-methoxybenzoic acid
Dichlobenil	2,6-Dichlorobenzonitrile
Dieldrin	Hexachloro-epoxy-octahydro-*endo,exo*-dimethano-naphthalene
Difolatan	*cis*-N-((1,1,2,2,-Tetrachloroethyl)thio)-4-cyclohexene-1,2-dicarboximide
Dimilin	1-(4-Chlorophenyl) 3-(2,6-difluorobenzoyl urea (IUPAC)
Dinoseb	2,4-Dinitro-6-S-butylphenol
Diphenamid	N,N-Dimethyldiphenylacetamide
Diphenyl (Biphenyl)	—
Diquat	1,1'-Ethylene-2,2'-bipyridylium dibromide
Diuron	3-(3,4-Dichlorophenyl)-1,1-dimethylurea
2,4-DP	2-(2,4-Dichlorophenoxy) propionic acid
Dyfonate	O-Ethyl-S-phenylethyl phosphonodithionate
Eptam	S-Ethyl dipropylthiocarbamate
EPTC	S-Ethyl dipropylthiocarbamate

Table VII. (*Continued*)

Name used in text	Chemical designation
Fluometuron	1,1-Dimethyl-3-(α,α,α-trifluoro-4-methylphenyl)-1,1-dimethylurea
Glyphosate (Roundup)	Isopropylamine salt of N-(phosphonomethyl)glycine
Heptachlor	1,4,5,6,7,8,8-Heptachloro-3a,4,7,7a-tetrahydro-4,7-methanoindene
Ioxynil	4-Hydroxy-3,5-diiodobenzonitrile
Lindane	γ-1,2,3,4,5,6-Hexachlorocyclohexane
Linuron	3-(3,4-Dichlorophenyl)-1-methoxy-1-methylurea
Malathion	O,O-Dimethylphosphorodithioate of diethyl-mercaptosuccinate
Mancozeb	Zinc ion and manganese ethylene bisdithiocarbamate
MCPA	2-Methyl-4-chlorophenoxyacetic acid
Methabenzthiazuron	3-(2-Benzothiazolyl)-1,3-dimethylurea
Methoprene (Altosid)	Isopropyl (2 E-4E)-11-methoxy-3,7,11-trimethyl-2,4-dodecadienoate
Methoxychlor	1,1,1-Trichloro-2,2-bis(4-methoxyphenyl) ethane
2-Methyl-4,6-diniatrophenol	—
Metobromuron	3-(4-Bromophenyl)-1-methoxy 1-methylurea
Metribuzin	3-Methylthio-4-amino-6-*tert*-butyl-1,2,4-triazine-5(4H)-one
Mirex	Dodecachlorooctahydro-1,3,4-metheno-2H-cyclobuta-(*cd*)pentalene
Monuron	3-(4-Chlorophenyl)-1,1-dimethylurea
Neburon	1-Butyl-3-(3,4-dichlorophenyl)-1-methylurea
Paraquat	1,1'-Dimethyl-4,4'-bipyridylium dichloride
Parathion	O,O-Diethyl O-p-nitrophenyl phosphorothioate
Pentachlorophenol (PCP)	—
Permethrin	3-(Phenoxyphenyl)methyl ($\pm$)-*cis, trans*-3-(2,2-dichloroethenyl)-2,2-dimethyl cyclopropane-carboxylate
Phaltan	N-(Trichloromethylthio)phthalimide
Phorate	O,O-Diethyl S-[(ethylthio)methyl] phos-phorodithioate
Picloram	4-Amino-3,5,6-trichloropicolinic acid
Prometryne	2-Methylthio-4,6-bis(isopropylamino)-S-triazine
Propanil	3,4-Dichloropropionanilide
Propazine	2-Chloro-4,6-bis(isopropylamino)-S-triazine
Propham	N-Phenyl isopropylcarbamate
Pydrin (Fenvalerate)	Cyano(3-phenoxyphenyl)methyl 4-chloro-*alpha*-(1-methylethyl)benzeneacetate
Rotenone	—
Siduron	3-Phenylurea-1-(2-methylcyclophexyl)
Simazine	2-Chloro-4,6-bis(ethylamino)-S-triazine
Solan	N-(3-Chloro-4-methylphenyl)-2-methylpentanamide
2,4,5-T	2,4,5-Trichlorophenoxyacetic acid
TCA	Trichloroacetic acid
TEPP	Tetraethyl diphosphate
Toxaphene	Chlorinated camphene (content of combined chlorine, 67–69%)
Trifluralin	2,6-Dinitro-N,N-dipropyl-4-trifluoromethylaniline
Vapam (Metam-Sodium)	Sodium-N-methyldithiocarbamate

Bioassay as a monitoring tool 125

AUDUS, L. J.: Plan growth substances. New York: Interscience (1965).
BANN, J. M., T. J. DECINO, N. W. EARLE, and Y. P. SUN: The fate of aldrin and dieldrin in the animal body. J. Agr. Food Chem. 4, 937 (1956).
BARKAI-GOLAN, R., and F. S. LATTAR: The effect of captan on rot-causing fungi in vitro. Isr. J. Agr. Res. 12, 131 (1962).
BATES, A. N., D. M. SPENCER, and R. L. WAIN: Investigations on fungicides. V. The fungicidal properties of 2-methyl 4,6 di-nitrophenol and some of its esters. Ann. Applied Biol. 50, 21 (1962).
BEROZA, M.: Current usage and some recent developments with insect attractans and repellents in the USDA. In M. Beroza (ed.): Chemicals controlling insect behavior, pp. 145–163. New York: Academic Press (1970).
BEWLEY, J. D., and M. BLACK: Physiology and biochemistry of seeds. New York: Springer-Verlag (1978).
BUSVINE, J. R.: A critical review of the techniques for testing insecticides. 2nd Ed. London: Commonwealth Agricultural Bureaux (1971).
CASIDA, J. E.: Sulfoxidation of thiocarbamate herbicides and metabolism of thiocarbamate sulfoxides in living mice and liver enzyme systems. Pest. Biochem. Physiol. 5, 1 (1975).
——, R. A. GRAY, and H. TILLES: Thiocarbamate sulfoxides: potent, selective and biodegradable herbicides. Science 184, 573 (1974).
Chevron Chemical Co.: Internal publication.
CORBETT, J. R.: The biochemical mode of action of pesticides. London: Academic Press (1974).
COX, H. W., and G. D. BOARDMAN: Growth response of a blue-green algae to 2,4-D and 2,4,5-T. Nat. Conf. Environ. Eng. Spec. Conf., p. 166 (1978).
DA SILVA, J. F., R. O. FADAYOMI, and G. F. WARREN: Cotyledon disc bioassay for certain herbicides. Weed Sci. 24, 250 (1976).
DEWEY, J. E.: Utility of bioassay in the determination of pesticide residues. J. Agr. Food Chem. 6, 274 (1958).
DICKSON, T. L., J. CAIRNS, JR., and R. J. LIVINGSTON (ed.): Biological data in water pollution assessment: Quantitative and statistical analysis. Philadelphia: ASTM STP 652 (1977).
EBERLE, D. O., and H. R. GERBER: Comparative studies of instrumental and bioassay methods for the analysis of herbicide residues. Archives Environ. Contam. Toxicol. 4, 101 (1976).
EDWARDS, C. A., S. D. BECK, and E. P. LICHTENSTEIN: Bioassay of aldrin and lindane in soil. J. Econ. Entomol. 51, 380 (1957).
Environmental Protection Agency: Proposed guidelines. Fed. Reg. 43 (132), pp. 29734 thru 29737 (1978).
—— Manual of biological testing methods for pesticides and devices. EPA Office of Pesticide Programs, Washington, D. C. (1973).
ESHEL, Y., and G. G. WARREN: A simplified method for determining phytotoxicity, leaching and adsorption of herbicides in soil. Weeds, 15, 115 (1967).
ETO, M.: Organophosphorus pesticides: Organic and biological chemistry. Cleveland: CRC Press (1974).
FRICK, E. L.: Methods of reducing variability in the results of glass slide spore germination assays of fungitoxicity. Ann. Applied Biol. 54, 349 (1964).
GAMBORG, O. L.: Plant tissue culture. Nat. Res. Council Canada, Ottawa (1975).
GEORGOPOULOS, S. G., and V. E. VOMVOYANNI: Differential sensitivity of duphenyl-tolerant and dupenyl-sensitive strains of fungi to chlorinated nitrobenzenes and to some duphenyl derivatives. Can. J. Bot. 43, 765 (1965).
GLASS, G. E. (ed.): Bioassay techniques and environmental chemistry. Ann Arbor: Ann Arbor Science Publishers (1973).
GREAVES, M. P., S. L. COOPER, H. A. DAVIES, J. A. MARSH, and G. I. WINGFIELD: Methods of analysis for determining the effects of herbicides on soil microorganisms and their activities. Tech. Report No. 45, Agr. Res. Council. Weed. Res. Org. Oxford, UK (1978).

126 G. K. KOHN

GUNTHER, F. A., and R. C. BLINN: Analysis of insecticides and acaracides. New York: Interscience (1955).

HARRIS, C. R., and E. P. LICHTENSTEIN: Factors affecting the volatilization of insecticidal residues in soils. J. Econ. Entomol. 54, 1038 (1961).

HARRIS, R. L., W. F. CHAMBERLAIN, and E. D. FRAZAR: Horn flies and stable flies: Free-choice feeding of methoprene mineral blocks to cattle for control. J. Econ. Entomol. 63, 384 (1974).

HELLING, C. S., D. D. KAUFMAN, and C. T. DIETER: Algae bioassay detection of pesticide mobility in soils. Weed Sci. 19, 685 (1974).

HOROWITZ, M.: Application of bioassay techniques to herbicide investigations. Weed Res. 16, 209 (1976).

HORSEFALL, J. G.: Fungicides and their action. Waltham, Mass.: Chronica Botanica (1945).

—— Principles of fungicidal action. Waltham, Mass.: Chronica Botanica (1956).

HOSKINS, W. M., and P. S. MESSENGER: Bioassay method: Musca. Adv. Chem. Ser. 1, 93 (1950).

INSCOE, M. N., and M BEROZA: Analysis of pheromones and other compounds controlling insect behavior. In G. Zweig (ed.): Analytical methods for pesticides and plant growth regulators VIII, 31 et seq. New York, London: Academic Press (1976).

JUSTICE, O. L., and L. N. BASS: Principles and practice of seed storage. Agr. Handbook No. 506, U.S. Gov't. Printing Office, Washington, D. C. (1978).

KAUFMAN, D. D., G. G. STILL, E. D. PAULSON, and S. K. BANDAL: Bound and conjugated pesticide residues. Amer. Chem. Soc. Symposium Ser. 29, (1976).

KELMAN, A.: Sourcebook of laboratory exercises in plant pathology. San Francisco: Freeman (1967).

KOCHER, C., W. ROTH, and J. TREBAUX: Bioassay tests: Daphnia. Mitt. Schweiz. Entomol. Ges. 26, 47 (1953).

KOHN, G. K.: The sulfenimide fungicides. In J. R. Plimmer (ed.): Pesticide chemistry in the 20th century. Amer. Chem. Soc. Symposium Ser. 37 (1977).

—— The pesticide industry. In J. A. Kent (ed.): Riegel's handbook of industrial chemistry, 7th Ed. New York: Van Nostrand (1974).

KRATKY, B. A., and G. F. WARREN: The use of three simple, rapid bioassays on forty-two herbicides. Weed Res. 11, 257 (1971).

LAUG, E. P., and O. G. FITZHUGH: 2,2-Bis-(p-chlorophenyl)-1,1,1-trichloroethane (DDT) in the tissues of the rat following oral ingestion for periods of six months to two years. J. Pharmacol. Exp. Therap. 87, 18 (1946).

LIANG, T. T., and E. P. LICHTENSTEIN: Synergism of insecticides by herbicides: Effect of environmental factors. Science 186, 1128 (1974).

—— —— Effect of light, temperature and pH on the degradation of azinophosmethyl. J. Econ. Entomol. 65, 315 (1972).

LICHTENSTEIN, E. P. and J. R. CASIDA: Myristicin, an insecticide and synergist occurring naturally in the edifice parts of parsnips. J. Agr. Food Chem. 11, 401 (1963).

——, T. T. LIANG, and T. W. FUHREMANN: A compartmentalized microcosm for studying the fate of chemicals in the environment. J. Agr. Food Chem. 26, 945 (1978).

—— ——, K. R. SCHULZ, H. K. SCHNOES, and G. T. CARTER: Insecticidal and synergistic components isolated from dill plants. J. Agr. Food Chem. 22, 658 (1974).

——, ——, T. W. FUHREMANN, and K. R. SCHULZ: Translocation and metabolism of [^{14}C] phorate as affected by percolating water in a model soil-plant ecosystem. J. Agr. Food Chem. 22, 991 (1974).

——, T. T. LIANG, and B. N. ANDEREGG: Synergism of insecticides by herbicides. Science, 181, 847 (1973).

——, D. G. MORAN, and C. H. MUELLER: Naturally occurring insecticides in cruciferous crops. J. Agr. Food Chem. 12, 158 (1963).

——, G. R. Mydral, and K. R. Schulz: Absorption of insecticidal residues from contaminated soils into five carrot varieties. J. Agr. Food Chem. 13, 126 (1965).
——, K. R. Schulz, T. W. Fuhremann, and T. T. Liang: Biological interaction between plasticizers and insecticides. J. Econ. Entomol. 62, 761 (1969).
—— ——, R. F. Skrentny, and Y. Tsukano: Toxicity and fate of insecticide residues in water. Arch. Environ. Health 12, 199 (1966).
—— —— Translocation of some chlorinated hydrocarbon insecticides into the aerial parts of pea plants. J. Agr. Food Chem. 8, 452 (1960).
—— —— ——, and P. A. Stitt: Insecticidal residues in cucumbers and alfalfa grown in Aldrin- or Heptachlor-treated soils. J. Econ. Entomol. 58, 742 (1965).
—— ——, and G. T. Cowley: Inhibition of the conversion of Aldrin to Daldrin in soils with methylenedroxyphenyl synergists. J. Econ. Entomol. 56, 485 (1963).
——, F. M. Strong, and D. G. Morgan: Identification of 2-phenethyl isothiocyanate as an insecticide occurring naturally in the edible parts of turnips. J. Agr. Food Chem. 10, 30 (1962).
——, L. J. Depew, E. L. Eshbaugh, and J. P. Sleesman: Persistence of DDT, aldrin and lindane in some Midwestern soils. J. Econ. Entomol. 53, 136 (1960).
Lukens, R. N.: Controlled release of foliar protectants. Devel. Ind. Microbiol. 16, 333 (1975).
——, and S. H. Ou: Chlorthalonil residues in field tomatoes and protection against alternaria solani. Phytopathol. 66, 1018 (1976).
Lukens, R. J.: Chemistry of fungicidal action. London: Chapman & Hall (1971).
Matsumura, F.: Toxicology of insecticides, p. 43. New York: Plenum Press (1975).
Matsunaka, S.: Propanil hydrolysis: Inhibition in rice plants by insecticides. Science 160, 1360 (1968).
McCann, A. E., and D. R. Cullimore: Influence of pesticides on the soil algal flora. Residue Reviews 72, 1 (1979).
Menn, J. J.: Comparative aspects of pesticide metabolism in plants and animals. Environ. Health Perspectives 27, 113 (1978).
——, E. Benjamini, and W. M. Hoskins: The effects of temperature and stage of life cycle upon the toxicity and metabolism of DDT in the housefly. J. Econ. Entomol. 50, 67 (1957).
Metcalf, R. L.: Organic insecticides, their chemistry and mode of action. New York: Interscience (1955).
—— A model ecosystem for the evaluation of biodegradability and ecological magnification. Outlook on Agr. 7, 55 (1972).
——, and R. B. March: Further studies on the mode of action of organic thionophosphate insecticides. Ann. Entomol. Soc. Amer. 46, 63 (1953).
——, T. R. Fukuto, and R. B. March: Plant metabolism of dithiosystox and thimet. J. Econ. Entomol. 50, 338 (1957).
Mitchell, J. E.: Bioassay of fungicide on treated seed. In A. Kelman (Chmn. Ed. Bd.): Sourcebook of laboratory exercises in plant pathology, p. 311. San Francisco: Freeman (1967).
Mitchell, J. W., and G. A. Livingston: Methods of studying plant hormones and growth-regulating substances. Agr. Handbook No. 336, Agr. Res. Serv. USDA, Washington, D. C. (1968).
Munnecke, D. E.: Estimation of foliar fungicide residues. In A. Kelman (Chmn. Ed. Bd.): Sourcebook of laboratory exercises in plant pathology, p. 326. San Francisco: Freeman (1967).
Neely, D.: The value of in vitro fungicide tests. Ill. Nat. Hist. Survey, Biol. Notes No. 64, Urbana (1969).
Nolan, K., and F. Wilcoxon: Method of bioassay for traces of parathion in plant material. Agr. Chemicals 5, 53 (1950).
Noll, M., and U. Bauer: Rapid sensitive herbicide bioassay by inhibition of trichome-migration of blue-green algae. Zbl. Bakt. Hyg. 1 Abt. Orig. B 157, 178 (1973).

O'BRIEN, M. C., and G. N. PRENDEVILLE: A rapid sensitive bioassay for the determination of paraquat and diquat in water. Weed Res. 18, 301 (1978).
PAGAN, C., and R. M. HAGEMAN: Determination of DDT by bioassay. Science 112, 222 (1950).
PALLOS, F. M., and J. E. CASIDA (ed.): The chemistry and action of herbicide antidotes. New York: Academic Press (1978).
PARKER, C.: The importance of shoot entry in the action of herbicides applied to the soil. Weeds 14, 117 (1966).
PILLAY, A. R., and Y. T. TCHAN: Study of soil algae. VII. Adsorption of herbicides in soil and prediction of their rate of application by algae methods. Plant and Soil 36, 571 (1972).
Plant Growth Regulator Working Group: Plant growth regulator handbook, Longmont, CO: The Group (1977).
PLAPP, F. W., JR.: Polychlorinated biphenyl: An environmental contaminant acts as an insecticide synergist. Environ. Entomol. 1, 580 (1972).
READY, D., and V. Q. GRANT: A rapid sensitive method for determination of 2,4-D acid in aqueous solution. Bot. Gaz. 109, 39 (1947).
REID, C. P. P., and W. HURTT: A rapid bioassay for simultaneous identification and quantitation of Picloram in aqueous solution. Weed Res. 9, 136 (1969).
RILEY, D., W. WILKINSON, and B. J. TUCKER: In D. D. Kaufman (ed.): Bound and conjugated pesticide residues. Amer. Chem. Soc. Symposium Ser. 29 (1976).
ROELOF:, W. L.: Threshhold hypothesis for pheromone perception. J. Chem. Ecol. 4, 685 (1978).
SAGGERS, D. T.: The search for new herbicides. In: Herbicide biochemistry, physiology and ecology, p. 447. New York: Academic Press (1976).
SCHAEFFER, C. H., and W. H. WILDER: Insect development inhibitors: A practical evaluation as mosquito control agents. J. Econ. Entomol. 65, 1066 (1972).
SCHECHTER, M. S., and I. HORNSTEIN: Chemical analysis of pesticide residues. Pest Control Res. I, 353 (1957).
SCHULZ, K. R., E. P. LICHTENSTEIN, T. T. LIANG, and T. W. FUHREMANN: Persistence and degradation of azinphosmethyl in soils, as affected by formulation and mode of application. J. Econ. Entomol. 63, 432 (1970).
SCOPES, N. E. A., and E. P. LICHTENSTEIN: The use of Folsomia fimetaria and Drosophila melanogasta as test insects for the detection of insecticide residues. J. Econ. Entomol. 60, 1539 (1967).
SHENNAN, J. L., and W. W. FLETCHER: The growth in vitro of micro-organisms in the presence of substituted phenoxyacetic and phenoxybuyric acid. Weed Res. 5, 266 (1965).
SIMS, J. J., H. MEE, and D. C. ERWIN: Methyl 2-benzimidazole carbamate, a fungitoxic compound isolated from cotton plants treated with methyl 1-(butylcarbamoyl)-2-benzimidazole-carbamate (benomyl). Phytopathol. 59, 1775 (1969).
SRI International: Aquatic organisms procedure; 333 Ravenswood Avenue, Menlo Park, CA 94025.
SUN, T. Y. T., and Y. P. SUN: Microbioassay of insecticides in milk by a feeding method. J. Econ. Entomol. 46, 927–930 (1953).
SUN, Y. P.: Bioassay insects. In G. Zweig (ed.): Analytical methods for pesticides, plant growth regulators, and food additives, vol. I, p. 399. New York: Academic Press (1963).
—— Bioassay of pesticide residues. Adv. Pest Control Res. I, 444 (1957).
——, and J. E. PANKASKIE: Drosophila, a sensitive insect, for the microbioassay of insecticide residues. J. Econ. Entomol. 47, 180 (1954).
TCHAN, Y. T., and C. M. CHIOU: Bioassay of herbicides by bioluminescence. ACTA Phytopath. Acad. Sci. Hungarica 12, 3 (1977).
——, J. E. ROSEBY, and G. R. FUNNELL: A new rapid specific bioassay method for photosynthesis inhibiting herbicides. Soil Biol. Biochem. 7, 39 (1975).
TERRIERE, L. C., and D. W. INGALSBE: Translocation and residual action of soil insecticides. J. Econ. Entomol. 46, 751 (1953).

THOMPSON, C. R., and G. KATS: Effects of H₂S fumigation on crop and forest plants. Environ. Science and Technol. **12**, 550 (1978).

TORGESON, D. C. (ed.): Fungicides, an advanced treatise, vol. I *et seq.* New York Academic Press (1969).

TRUELOVE, B. (ed): Research methods in weed science; 2nd ed. Auburn: S. Weed Sci. Soc. (1977).

——, D. E. DAVIS, and L. R. JONES: A new method for detecting photosynthesis inhibitors. Weed Sci. **22**, 15 (1974).

TSAO, P. H.: A serial dilution endpoint method for estimating disease potentials of citrus phytophoras in soil. Phytophathol. **50**, 717 (1960).

Union Carbide Corporation: Aquatic environmental sciences. Sand Mill River Road 100C, Tarrytown, New York 10591.

WASSERBURGER, H. J.: Daphnia magna as a test animal for the determination of traces of contact insecticides. Pharmazie **7**, 731 (1952).

WEAVER, R. J.: Plant growth substances in agriculture. San Francisco: Freeman (1972).

WRIGHT, S. J. L.: A simple agar plate method for herbicide bioassay or detection. Bull. Environ. Contam. Toxicol. **14**, 65 (1975).

ZILKAH, S., and J. GRESSEL: Cell cultures versus whole plants for measuring phytotoxicity. I. The establishment and growth of callus and suspension cultures; definition of factors affecting toxicity on calli. Plant and Cell Physiol. **18**, 641 (1977).

—— —— Cell cultures versus whole plants for measuring phytotoxicity. III. Correlation between phytotoxicity in cell suspension cultures, calli, and seedlings. Plant and Cell Physiol. **18**, 815 (1977).

——, P. E. BOCION, and J. GRESSEL: Cell cultures versus whole plants for measuring phytotoxicity. II. Correlation between phytotoxicity in seedling and calli. Plant and Cell Physiol. **18**, 657 (1977).

Zoecon Corporation: Internal data.

ZWEIG, G., J. E. HITT, and R. McMAHON: Effect of certain quinones, diquat and diuron on *Chlorella pyrenordosa* chick (Emerson strein). Weed Sci. **6**, 69 (1968).

—— Analytical methods for pesticides, plant growth regulators, and food additives. New York: Academic Press (1963 to present).

Manuscript received May 27, 1979; accepted January 23, 1980.

Insecticide resistance and prospects for its management

By

George P. Georghiou[*]

Contents

I. Introduction

After a decade of relative neglect, resistance to insecticides[1] is again beginning to receive the attention it deserves as a challenging problem that requires new solutions.

As reflected in the professional literature, the late 1960s and early 1970s have witnessed a decline in research on resistance (Fig. 1) with an almost fatalistic resignation to the inevitability of its occurrence. Even industry had been increasingly hesitant to develop new chemicals for commercial use, as evident in the substantial decline in the number

[*] Division of Toxicology and Physiology, Department of Entomology, University of California, Riverside, CA 92521. Presented in part at the April 1979 US-ROC Cooperative Science Program seminar on "Environmental Problems Associated with Pesticide Usage in the Intensive Agricultural System," Taipei, Taiwan, Republic of China, as sponsored by the National Science Foundation (U.S.A.) and the National Science Council (R.O.C.).

[1] Common, trade, and chemical names of the pesticides discussed in this paper are presented in Table I.

132 G. P. Georghiou

Table I. *Chemical designations of pesticides mentioned in text.*

Pesticide[a]	Chemical designation
bromophenothrin	3-phenoxybenzyl (1R)-cis-(2,2-dibromovinyl)-2,2-dimethylcyclopropanecarboxylate (RU 23603)
chlorpyrifos	O,O-diethyl O-(3,5,6-trichloro-2-pyridyl) phosphorothioate
chlorpyrifos-methyl	O,O-dimethyl O-(3,5,6-trichloro-2-pyridyl) phosphorothioate
cismethrin	5-benzyl-3-furylmethyl (1R)-cis-3-(2,2-dimethylvinyl)-2,2-dimethylcyclopropanecarboxylate (NRDC 119)
cypermethrin	(R,S)-α-cyano-3-phenoxybenzyl (1R)-trans-3-(2,2-dichlorovinyl)-2,2-dimethylcyclopropanecarboxylate (RU 24298)
decamethrin	(S)-α-cyano-3-phenoxybenzyl (1R)-cis-3-(2,2-dibromovinyl)-2,2-dimethylcyclopropanecarboxylate (NRDC 161)
DEF	S,S,S-tributylphosphorotrithioate
diflubenzuron	1-(4-chlorophenyl)-3-(2,6-difluorobenzoyl)-urea
dimethoate	O,O-dimethyl S-(N-methylcarbamoylmethyl) phosphorodithioate
fenitrothion	O,O-dimethyl O-(4-nitro-m-tolyl) phosphorothioate
fenthion	O,O-dimethyl O-[4-(methylthio)-m-tolyl] phosphorothioate
fenvalerate	(R,S)-α-cyano-3-phenoxybenzyl (R,S)-2-(p-chlorophenyl)-3-methylbutyrate (S-5602, Pydrin)
methoprene	isopropyl 11-methoxy,3-7,11-trimethyldodeca-2,4-dienoate
parathion	O-O-diethyl O-p-nitrophenyl phosphorothioate
parathion-methyl	O,O-dimethyl O-p-nitrophenyl phosphorothioate
(1R)-cis-permethrin	3-phenoxybenzyl (1R)-cis-3-(2,2-dichlorovinyl)-2,2-dimethylcyclopropanecarboxylate (NRDC 167)
(1R)-trans-permethrin	3-phenoxybenzyl (1R)-trans-3-(2,2-dichlorovinyl)-2,2-dimethylcyclopropanecarboxylate (NRDC 147)
piperonyl butoxide	α-[2-(2-butoxyethoxy)ethoxy]-4,5-methylenedioxy-2-propyltoluene
temephos	O-O-dimethyl phosphorothioate, O,O-diester with 4,4-thiodiphenol

[a] Trade names are provided for identification only. No endorsement is implied.

of compounds submitted to the World Health Organization for testing (Fig. 2). Yet, as various approaches to pest control are being reappraised and programs of integrated pest management (IPM) are being formulated, it is becoming obvious that pesticides will continue to be an essential means of pest control in the great majority of situations.

The revived interest in studies of resistance that we are now witnessing is directed toward the development of resistance-delaying or resistance-avoiding tactics in the context of IPM. Such research is further prompted by the realization that in order for an IPM program to remain viable it must ensure that resistance does not evolve or that it is at least sensibly forestalled. This is because an imposed change to another pesticide due to resistance could be so disruptive to the biological control component of IPM that an entirely new program would then need to be formulated. Such programs require prolonged experimentation.

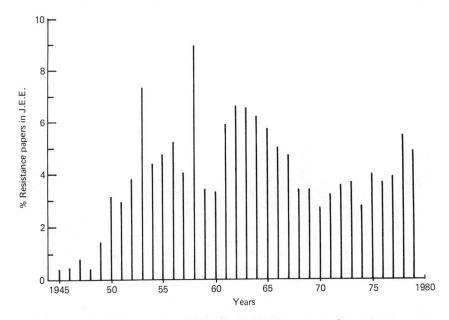

Fig. 1. Percentage of papers concerned directly with insecticide resistance among articles published in the Journal of Economic Entomology (J.E.E.) during 1945 to 1979.

Thus, research on resistance is again increasing, but with a sensibly different orientation; whereas in the past the challenge of continued effective pest control rested almost solely with the synthetic chemist, now it must be shared at least equally by the pest management specialist.

II. Status of resistance

The number of species of insects and mites in which cases of resistance were reported through 1975 totaled 364 (GEORGHIOU and TAYLOR 1976). The available records have been updated through 1978 and are summarized in Table II according to the taxonomic order of the species, whether of medical or agricultural importance, and the chemical class of the insecticide involved. It is evident that resistance now involves at least 414 species of insects and mites and that the most significant increases during the last few years have occurred in the species of agricultural importance, especially in the lepidoptera and in the acarina.

However, the number of resistant species alone does not tell the entire story. One should also take into account the large number of chemicals that many resistant strains can now tolerate, the expanding geographical distribution of resistant populations, and the frequency of the resistant genes. For example, resistance in the mosquitoes *Aedes nigromaculis* and

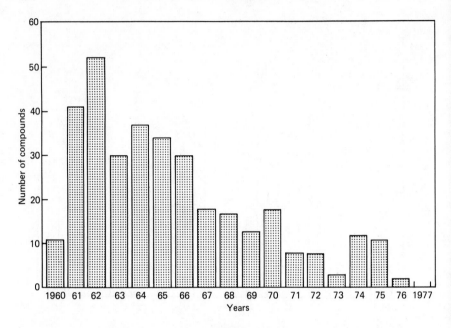

Fig. 2. Yearly numbers of compounds cleared through Stage I of the insecticide testing
program of the World Health Organization during 1960 to 1977 (WHO 1977).

Table II. *Number of species of Arthropoda with reported cases of
resistance to pesticides.*[a]

Order	Pesticide group[b]					Importance[c]		Total
	DDT	Cyclod.	OP	Carb.	Other	Med.	Agr.	
Acarina	17	15	41	6	28	15	38	53
Anoplura	4	3	2	1	—	5	—	5
Coleoptera	24	56	28	10	21	—	64	64
Dermaptera	1	—	—	—	—	—	1	1
Diptera	91	100	50	7	5	115	27	142
Ephemeroptera	2	—	—	—	—	—	2	2
Hemipt./Het.	7	15	5	—	—	5	13	18
Hemipt./Hom.	13	12	29	7	5	—	43	43
Hymenoptera	1	2	—	—	—	—	3	3
Lepidoptera	41	40	31	15	4	—	64	64
Mallophaga	—	3	—	—	—	3	—	3
Orthoptera	2	3	1	1	1	3	—	3
Siphonaptera	6	5	1	—	—	6	—	6
Thysanoptera	3	5	1	—	2	—	7	7
Totals	212	259	189	47	66	152	262	414

[a] Updated from GEORGHIOU and TAYLOR (1976).
[b] Cyclod. = cyclodiene, OP = organophosphate, Carb. = carbamate.
[c] Med. = medical, Agr. = agricultural.

Culex quinquefasciatus in California now involves nearly all insecticides that are registered for their control (GEORGHIOU *et al.* 1975, SCHAEFER and WILDER 1970). Resistance to dimethoate in the green peach aphid in Britain is so generally distributed that of 258 collections that were examined, only 3 did not contain resistant individuals. With 196 collections, more than 76% of the aphids were resistant (SAWICKI *et al.* 1978). The frequency of DDT-resistance genes in *Anopheles culicifacies* in India during 1970 to 1971 was calculated to have been 0.34. In *Anopheles albimanus* in certain areas of El Salvador, the frequency of DDT- and propoxur-resistance genes during 1970 to 1972 was found to be 0.80 and 0.48, respectively (GEORGHIOU and TAYLOR 1976). Such extremely high frequency of resistance genes in mosquitoes is usually encountered in intensely agricultural environments where the mosquito population is exposed to indirect selection by sprays applied to crops.

Despite the severity of the problem of resistance in several pests, chemical control is still being achieved in the majority of cases, and one must credit the synthetic chemist for a number of notable advances that have been made in recent years, including the discovery of new organophosphates, derivatized carbamates, insect growth regulators, and synthetic pyrethroids. It is realized, however, that resistance to these chemicals can also develop under appropriate conditions, as has already been demonstrated in the laboratory. For example, a high degree of resistance to pyrethroid insecticides has been induced in larvae of *Culex quinquefasciatus* after 10 to 20 generations of selection pressure (PRIESTER and GEORGHIOU 1978). The degree of resistance toward the selecting agents, *trans*- and *cis*-permethrin, amounted to 4100x and 450x, respectively, while cross resistance extended to nearly every pyrethroid tested, included fenvalerate (*ca.* 6000x), bromophenothrin (>1,000x), cypermethrin (432x), decamethrin (142x), and cismethrin (47x) (PRIESTER and GEORGHIOU 1979). At least in mosquitoes and house flies, a large part of pyrethroid resistance is due to one of the components of DDT resistance, the gene *kdr* (FARNHAM 1977, PRIESTER and GEORGHIOU 1979). DDT resistance was observed to rise to high levels concurrently with pyrethroid resistance during selection by pyrethroids (PRIESTER and GEORGHIOU 1979). Likewise, pyrethroid resistance was found to increase during selection by DDT (FARNHAM and SAWICKI 1976, OMER *et al.* 1979).

In addition to the pyrethroids, induction of resistance has been demonstrated toward the juvenile hormone mimic, methoprene, in the house fly (GEORGHIOU *et al.* 1978), mosquitoes (BROWN and BROWN 1974, GEORGHIOU *et al.* 1974) and various other species of insects (BROWN *et al.* 1978), and toward the chitin synthesis inhibitor, diflubenzuron, in the house fly (PIMPRIKAR and GEORGHIOU 1979).

It is obvious that the risk of resistance will continue to exist regardless of the chemical nature of the pesticide being used; therefore, new approaches to the problem are urgently needed. There are two general areas in which research should prove profitable: First, we must develop the capability of quantifying the risk for resistance in a given situation. Second, we must refine our specifications for pesticide usage, including thresholds for treatment, types of formulation, and choice of chemical, so that the selection coefficient for resistance is substantially reduced. I will attempt to discuss these areas briefly below.

III. Improving predictive capability

During the last few years, several papers have been published on the dynamics of resistance based primarily on computer simulations. Research by various groups in England (CURTIS et al. 1978, COMINS 1977, 1978, WOOD and COOK 1978), Switzerland (MUIR 1977), Israel (GRESSEL and SEGEL 1978), and the United States (GEORGHIOU and TAYLOR 1977 a and b, HUETH and REGEV 1974, REGEV et al. 1979, TAYLOR and GEORGHIOU 1979 a and b, TAYLOR and HEADLEY 1975) has contributed significantly to our understanding of the selection process.

We now recognize three types of factors that influence the evolution of resistance: genetic, biological, and operational (Table III). The factors in the genetic and biological categories are inherent qualities of the species. Therefore, they are beyond our control, but their assessment is essential in determining the "risk for resistance" in a given situation. The operational factors, on the other hand, are man-made and are thus within our discretion. These factors can be altered to the extent necessary, depending on the risk for resistance that is revealed by the genetic and biological factors. Since a discussion of these factors has already been published (GEORGHIOU and TAYLOR 1976, 1977 a and b), I will only stress those that are most important under typical field situations. They include the *dominance* of the R *allele, dosage* of insecticide, *generation turnover, mobility* of the population, and *persistence* of residues. We have obtained quantitative estimates for each parameter with a computer model, which in its simplest form enables the calculation of changes in R gene frequency and population growth in successive generations based on the relative survival and reproductive potential of the three genotypes, SS, SR, and RR.

a) Dominance of R allele and insecticide dosage

Resistant populations evolve faster if resistance is dominant, slower if it is recessive. However, the expression of dominance is dependent upon the dose applied (CURTIS et al. 1978, TAYLOR and GEORGHIOU 1979 a). The regression lines that are presented in Figure 3 indicate the response of the three genotypes of *Culex quinquefasciatus* larvae (SS, RS, and RR) toward the pyrethroid insecticide, (1R)-*cis*-permethrin. When a small dose is applied (D_S), the heterozygotes survive, thus, the resistant allele is functionally dominant. Under these circumstances

Table III. *Known or suggested factors influencing the selection of resistance to insecticides in field populations.*[a]

A. Genetic
    1. Frequency of *R* alleles
    2. Number of *R* alleles
    3. Dominance of *R* alleles
    4. Penetrance; expressivity; interactions of *R* alleles
    5. Past selection by other chemicals
    6. Extent of integration of *R* genome with fitness factors
B. Biological
  a. Biotic
    1. Generation turn-over
    2. Offspring per generation
    3. Mongamy/polygamy; parthenogenesis
  b. Behavioral
    1. Isolation; mobility; migration
    2. Monophagy/polyphagy
    3. Fortuitous survival; refugia
C. Operational
  a. The chemical
    1. Chemical nature of pesticide
    2. Relationship to earlier used chemicals
    3. Persistence of residues; formulation
  b. The application
    1. Application threshold
    2. Selection threshold
    3. Life stage(s) selected
    4. Mode of application
    5. Space-limited selection
    6. Alternating selection

---

[a] Adapted from GEORGHIOU and TAYLOR (1976).

resistance evolves relatively fast. In contrast, if a large dose is applied (D_L), the heterozygotes are killed; thus, the resistant allele is functionally recessive, and resistance in this case evolves relatively slowly.

b) Generation turnover

The relation between generation turnover in various soil-inhabiting pest species and the number of years that elapsed until each species manifested resistance to soil applications of aldrin/dieldrin is shown in Figure 4. It will be seen that root maggots (*Hylemya* spp.), which complete 3 to 4 generations/yr, evolved resistance after 5 yr of exposure, while *Diabrotica longicornis*, with only 1 generation/yr, has required 8 to 10 yr to become resistant. *Popillia japonica* and *Amphimallon majalis*, also with 1 generation/yr, have required 9 to 14 yr for resistance. Finally, the sugarcane wireworm, *Melanotus tamsuyensis*, in Taiwan, which needs 2 yr to complete 1 generation, has taken 20 yr to develop resistance. It is evident that under these conditions, 10 to 15 generations are required on the average for the development of resistance.

138   G. P. GEORGHIOU

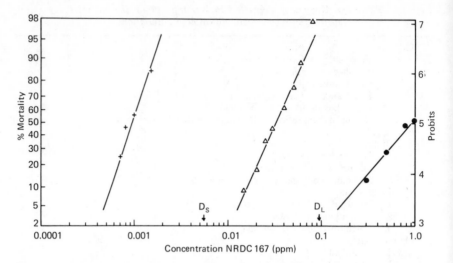

Fig. 3. Dose-response lines for larvae of *Culex quinquefasciatus* Say tested with (1R)-*cis*-permethrin (NRDC 167): +, susceptible (SS); Δ, heterozygous (RS); and •, resistant (RR). The dominance is seen to depend upon dose: with a small dose (D$_S$) resistance is functionally dominant, while with a large dose (D$_L$) it is functionally recessive. (Adapted from PRIESTER and GEORGHIOU 1978.)

c) Population mobility

An extremely important behavioral factor is population mobility and dispersal. The influx of migrants tends to dilute the frequency of resistance among survivors of treatments so that in otherwise comparable situations the evolution of resistance may be expected to be commensurate with the relative isolation of a population. By computer simulation, we have found that a moderate rate of immigration of susceptible individuals could ensure the containment of resistance if the initial population was of low density and if a short-lived pesticide was used in regular treatments (TAYLOR and GEORGHIOU 1979 a and b).

d) Persistence of residues

Since residues undergo decay, resistance that may be functionally recessive at the time of application eventually becomes functionally dominant as the residue is reduced below the threshold that is lethal to the heterozygotes. Figure 5 illustrates the significance of the rate of decay by comparing diagrammatically the duration of selection exerted on each genotype by compounds with different half-lives. It will be seen that a chemical with a half-life of 10 days may select for as long as 60 days, whereas one with a half-life of 1 day may select for only 6 days.

In agreement with expection, the simulation of selection by residues whose half-life varied from 1 to 10 days indicated that resistance evolves fastest with the most persistent chemical (TAYLOR and GEORGHIOU 1979 b). It is also significant to note that under the influence of migrants, resistance is delayed substantially, such delay being most pronounced when high dosages of the least persistent chemical are used.

e) Operational factors

The greatest opportunities for countering the phenomenon of resistance lie in our ability to limit the degree of selection pressure according to the propensity for resistance of the target population. This may be achieved by considering the operational factors that influence the evolution of resistance (Table III) and modifying them to the extent necessary. The importance of each of these factors is self-evident and may be easily documented with examples from the published literature.

It is apparent that resistance would be delayed the most, or may be avoided entirely, if the following conditions are fulfilled: (a) the pesticide has short chemical stability, (b) the pesticide is not related to an earlier used chemical with respect to mode of action or metabolism, (c) the formulation does not provide prolonged release of the chemical in the environment, (d) the selection threshold is relatively low, (e) selec-

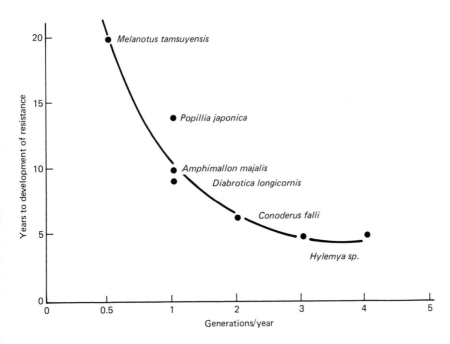

Fig. 4. Relationship between generation turnover and appearance of resistance in species selected by soil application of aldrin/dieldrin.

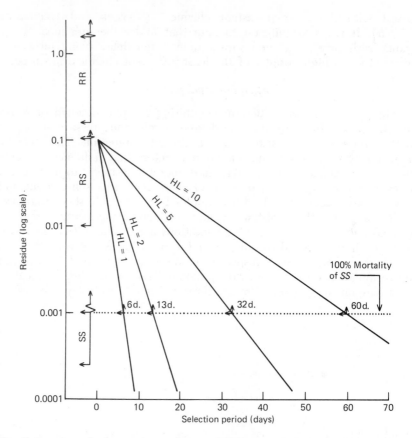

Fig. 5. Duration of selection by residues with different half-life (HL) properties. All residues are assumed to decay exponentially. Ranges of susceptibility of genotypes SS, RS, and RR approximate those for NRDC 167 in Figure 3. Selection is considered as ended when residues are reduced to LD_{50} level of SS.

tion is directed mainly against adults, (f) the application is localized rather than area-wide, and (g) certain generations are left untreated.

These measures may appear as extreme, and *in toto* they may be impracticable. But it should be remembered that the extent to which one or more of these measures may be necessary would depend on the "resistance risk" which has been ascribed to the target population.

IV. Chemical countermeasures

Chemical measures for countering resistance include the use of alternative insecticides, synergists, and the application of chemicals in mixtures or in rotations.

The practice most often followed when resistance appears is to increase the dosage of the chemical. Because higher dosages exert greater selection pressure, these increases in dosage soon prove inadequate. The chemical is subsequently applied more frequently, but with little or no benefit other than to eliminate susceptible individuals arriving from the outside. Thus, the application of a chemical at higher dosages or at greater frequency does not constitute a countermeasure for resistance that has already appeared.

The next usual move is to change to an alternative chemical. The chances for obtaining relatively lasting control would depend upon 2 factors: dissimilarity of detoxication pathways and dissimilarity of mode of action of the new chemical in relation to those employed previously. Nevertheless, if strong selection pressure is applied with no regard to the "operational" factors which have been mentioned earlier, resistance to the new compound will also develop eventually.

a) Synergists

A chemical countermeasure for resistance which is receiving increasing attention involves the joint use of an insecticide with a synergist. Synergists act by inhibiting specific detoxication enzymes and thus eliminate the selective advantage of individuals possessing such enzymes. An important consideration is that there should be no alternative resistance pathway available in the population.

Recently, we have demonstrated the application of this principle to the mosquito Culex quinquefasciatus (RANASINGHE and GEORGHIOU 1979). The strain that was studied possessed high resistance to several organophosphates. The use of the oxidase inhibitor, piperonyl butoxide (PB), in combination with any of these insecticides failed to produce synergism, thus indicating that this resistance was not due to an oxidase (Table IV). In contrast, treatment in combination with the esterase inhibitor, DEF®, reduced resistance almost to the level found in the susceptible strain (GEORGHIOU et al. 1978). We then showed by selection with temephos during 12 consecutive generations that resistance can be inhibited when the insecticide is used jointly with DEF®; but that it can advance to higher levels when the insecticide is used alone or in combination with PB (Table IV) (RANASINGHE and GEORGHIOU 1979).

The difficulty with synergists is that none of those that are presently available can be used under field conditions: DEF® is a defoliant and PB is unstable in sunlight. However, the principle is promising and we should see more progress in this direction in the near future.

b) Insecticide mixtures and rotations

Another approach, which is receiving renewed attention as a possible resistance-delaying tactic, is the use of insecticides in mixtures or

Table IV. *Changes in resistance levels*[a] *of* Culex quinquefasciatus
larvae under various selection regimes.[b]

Insecticide	Parental (Hanford) (Unselected)		Selected strains		
	F_1	F_{12}[c]	Temephos	Temephos + PB	Temephos + DEF®
Temephos	117	3.1	322	122	3.1
Chlorpyrifos	52	2.5	52	52	2.8
Methyl chlorpyrifos	83	—	197	250	2.5
Fenthion	49	4.9	87	66	11.1
Fenitrothion	12	—	32	22	6.5
Parathion	14	1.4	33	17	3.7
Methyl parathion	24	2.8	36	28	1.3

[a] LC_{50} resistant strain ÷ LC_{50} susceptible strain.
[b] Data from RANASINGHE (1976) and RANASINGHE and GEORGHIOU (1979).
[c] Resistance declined in the absence of insecticidal selection.

in rotations. Although this concept was introduced several years ago (CUTRIGHT 1959; reviews by BROWN 1968, 1970, and 1971), surprisingly little has been done to define the requirements for its application. However, some of the papers published on the subject suggest that where chemicals with contrasting modes of action and detoxication pathways are employed, a delay in the onset of resistance is noted (GRAVES *et al.* 1967, ASQUITH 1961, BURDEN *et al.* 1960, OZAKI *et al.* 1973). In the extreme case reported by PIMENTEL and BELLOTTI (1976), no resistance was obtained with selection by a mixture of 6 inorganic compounds, whereas positive results were obtained when each compound was used alone.

The principles of joint and sequential use of chemicals are currently being tested in our laboratory with 3 commercial insecticides that lack significant cross resistance, e.g. temephos, propoxur, and permethrin. A strain of *Culex quinquefasciatus* has been synthesized that contains genes for resistance to each of these compounds at the low frequency of 0.02. Subcolonies were then placed under selection pressure with each compound separately and in various combinations. After 10 generations each of the colonies that were selected by a single compound had developed high resistance to that compound but not to the other 2 chemicals. The colonies selected by combinations developed some resistance only toward propoxur when this carbamate was part of the combination (A. LAGUNES, still unpublished).

The concept of joint use of insecticides assumes that the mechanisms for resistance to each member chemical exist in such low frequency that they do not occur together in any single individual in the population. Thus, insects that may survive 1 of the chemicals are killed by the other. The concept of rotation assumes additionally that individuals that may be resistant to 1 chemical have substantially lower biotic fitness than the susceptible individuals, so that their frequency declines during the interval between applications of the same compound.

The possible oscillations in susceptibility of a population that is exposed to 4 chemicals used in rotation are presented diagrammatically in Figure 6: resistance to Compound A rises slightly in the generation in which it is applied and then declines during the subsequent 3 generations during which it is not applied. It rises in the 5th generation when the compound is again applied, but drops in generations 6, 7, and 8. The same pattern is anticipated for compounds B, C, and D, each chemical oscillating one step behind its predecessor in the treatment sequence.

We are entering an era of intensive research on the management of resistance. Emerging from this effort, in gradual fashion, will be the capability of accurately quantifying the risk for resistance that exists in a given situation and of reducing such risk through appropriate cultural and natural control practices and through greater sophistication in the choice and delivery of chemical control measures.

Summary

The incidence of resistance to pesticides has increased substantially in recent years, especially among arthropods of agricultural importance, while it has continued to present serious difficulties in the control of species of importance to public health. By the end of 1978 the number of species that have developed strains resistant to one or more pesticides had increased to 414. Furthermore, laboratory studies have demonstrated that resistance can also develop toward pesticides of novel mode of action such as juvenile hormone analogs (methoprene) and chitin synthesis inhibitors (diflubenzuron).

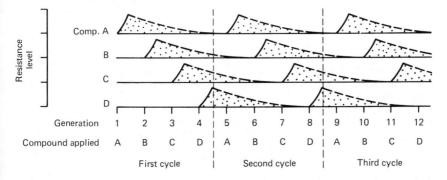

Fig. 6. Diagrammatic representation of expected oscillations in susceptibility of a population that is exposed to 4 unrelated chemicals (A–D) used in rotation against succeeding generations (1–12).

Increased awareness of actual and potential obstacles that resistance presents in the formulation of effective pest management programs has generated renewed interest in research on the dynamics and suppression of resistance. Emphasis is placed on the elucidation of the role of specific biological, behavioral, genetic, ecological, and operational parameters that influence resistance, and the utilization of this information in the formulation of strategies for its avoidance or suppression. The paper discusses the influence of a number of these parameters, including the role of gene dominance, insecticide dosage, residue persistence, and pest population dispersal. Operational considerations for forestalling resistance, including the use of synergists and the application of insecticides in suitable combinations or rotations, are also discussed.

References

ASQUITH, D.: Methods of delaying selection of acaricide-resistant strains of the European red mite. J. Econ. Entomol. **54**, 439 (1961).

BROWN, A. W. A.: Insect resistance. IV. Countermeasures for resistance. Farm Chemicals **127**, 50 (1970).

—— Insecticide resistance—Genetic implications and applications. World Rev. Pest Control **7**, 104 (1968).

—— Pest resistance to pesticides. In R. White-Stevens (ed.): Pesticides in the environment. Vol. I, p. 457 (1971).

BROWN, T. M., and A. W. A. BROWN: Experimental induction of resistance to a juvenile hormone mimic. J. Econ. Entomol. **67**, 799 (1974).

——, D. H. DeVRIES, and A. W. A. BROWN: Induction of resistance to insect growth regulators. J. Econ. Entomol. **71**, 223 (1978).

BURDEN, G. S., C. S. LOFGREN, and C. N. SMITH: Development of chlordane and malathion resistance in the German cockroach. J. Econ. Entomol. **53**, 1138 (1960).

COMINS, H. N.: The development of insecticide resistance in the presence of migration. J. Theor. Biol. **64**, 177 (1977).

—— The control of adaptable pests. In G. A. Norton and C. S. Holling (eds.): Pest management, p. 217. Pergamon Press (1979).

CURTIS, C. F., L. M. COOK, and R. J. WOOD: Selection for and against insecticide resistance and possible methods of inhibiting the evolution of resistance in mosquitoes. Ecol. Entomol. **3**, 273 (1978).

CUTRIGHT, C. R.: Rotational use of spray chemicals in insect and mite control. J. Econ. Entomol. **52**, 432 (1959).

FARNHAM, A. W.: Genetics of resistance of house flies (*Musca domestica* L.) to pyrethroids. I. Knockdown resistance. Pest. Sci. **8**, 631 (1977).

——, and R. M. SAWICKI: Development of resistance to pyrethroids in insects resistant to other insecticides. Pest. Sci. **7**, 278 (1976).

GEORGHIOU, G. P., and C. E. TAYLOR: Pesticide resistance as an evolutionary phenomenon. Proc. XV Internat. Congress Entomol., p. 759 (1976).

—— —— Genetic and biological influences in the evolution of insecticide resistance. J. Econ. Entomol. **70**, 319 (1977 a).

—— —— Operational influences in the evolution of insecticide resistance. J. Econ. Entomol. **70**, 653 (1977 b).

——, V. ARIARATNAM, M. PASTERNAK, and C. S. LIN: Organophosphate multiresistance in *Culex pipiens quinquefasciatus* in California. J. Econ. Entomol. **68**, 461 (1975).

——, S. LEE, and D. H. DeVRIES: Development of resistance to the juvenoid methoprene in the house fly. J. Econ. Entomol. **71**, 544 (1978).

——, C. S. Lin, and M. E. Pasternak: Assessment of potentiality of *Culex tarsalis* for development of resistance to carbamate insecticides and insect growth regulators. Proc. Pap. Calif. Mosq. Control Assoc. **42**, 117 (1974).

Graves, J. B., J. S. Roussel, J. Gibbens, and D. Patton: Laboratory studies on the development of resistance and cross-resistance in the boll weevil. J. Econ. Entomol. **60**, 47 (1967).

Gressel, J., and L. A. Segel: The paucity of plants evolving genetic resistance to herbicides: Possible reasons and implications. J. Theor. Biol. **75**, 349 (1978).

Hueth, D., and U. Regev: Optimal agricultural pest management with increasing pest resistance. Amer. J. Agr. Econ. **56**, 543 (1974).

Muir, D. A.: Genetic aspects of developing insecticide resistance of malaria vectors. Part II. Gene flow and control pattern. World Health Organization WHO/VBC/ 77.659, p. 10 (1977).

Omer, S. M., G. P. Georghiou, and S. N. Irving: DDT/pyrethroid resistance interrelationships in *Anopheles stephensi*. Mosquito News. Submitted (1979).

Ozaki, K., Y. Sasaki, M. Ueda, and T. Kassai: Results of the alternate selection with two insecticides and the continuous selection with mixtures of two or three ones of *Laodelphax striatellus* Fallen. Botyu-Kagaku **38**, 222 (1973).

Pimentel, D., and A. C. Bellotti: Parasite-host population systems and genetic stability. Amer. Nat. **110**, 877 (1976).

Pimprikar, G. D., and G. P. Georghiou: Mechanisms of resistance to diflubenzuron in the house fly, *Musca domestica* (L.). Pest. Biochem. Physiol. **12**, 10 (1979).

Priester, T. M., and G. P. Georghiou: Induction of high resistance to permethrin in *Culex pipiens quinquefasciatus*. J. Econ. Entomol. **71**, 197 (1978).

—— —— Cross-resistance spectrum in pyrethroid-resistant *Culex quinquefasciatus*. Pest. Sci. Accepted for publication (1979).

Ranasinghe, R. E.: Role of synergists in the selection of specific organophosphorus resistance mechanisms in *Culex pipiens quinquefasciatus* Say. Ph.D. Dissertation, Univ. Calif., Riverside, p. 122 (1976).

——, and G. P. Georghiou: Comparative modification of insecticide resistance spectrum of *Culex pipiens fatigans* Wied by selection with temephos and temephos/synergist combinations. Pest. Sci. **10**, 502 (1979).

Regev, W., H. Shalit, and A. P. Gutierez: Economic conflicts in plant protection: The problems of pesticide resistance; theory and application to the Egyptian alfalfa weevil. In G. A. Norton and C. S. Holling (eds.): Pest management, p. 281. New York: Pergamon Press (1979).

Sawicki, R. M., A. L. Devonshire, A. D. Rice, G. D. Moores, S. M. Petzing, and A. Cameron: The detection and distribution of organophosphorus and carbamate insecticide-resistant *Myzus persicae* (Sulz.) in Britain in 1976. Pest. Sci. **9**, 189 (1978).

Schaefer, C. H., and W. H. Wilder: Insecticide resistance and cross resistance in *Aedes nigromaculis*. J. Econ. Entomol. **63**, 1224 (1970).

Taylor, C. E., and G. P. Georghiou: Suppression of insecticide resistance by alteration of gene dominance and migration. J. Econ. Entomol. **72**, 105 (1979 a).

—— —— The influence of pesticide persistence in the evolution of resistance. J. Econ. Entomol. Submitted (1979 b).

Taylor, C. R., and J. C. Headley: Insecticide resistance and the evaluation of control strategies for an insect population. Can. Entomol. **107**, 237 (1975).

Wood, R. J., and L. M. Cook: Estimating selection pressures on insecticide resistance genes (Preliminary Note). World Health Organization WHO/VBC/78.683, p. 7 (1978).

World Health Organization: WHO program for evaluating and testing new insecticides. Summary 1975–77 tests. WHO/VBC/77.2, p. 3 (1977).

Manuscript received December 20, 1979; accepted January 24, 1980.

"Bound" residues in soils and transfer of soil residues in crops

By

E. PAUL LICHTENSTEIN *

Contents

I. Foreword

Data reported in this contribution are based on research efforts in our laboratory at the University of Wisconsin. References to the original publications of the various research topics are presented and indicate the contributions of B. A. ANDEREGG, T. W. FUHREMANN, J. KATAN, T. T. LIANG, K. R. SCHULZ and E. P. LICHTENSTEIN.

II. Problems of unextractable, bound insecticide residues in soils and their potential release

This section summarizes data obtained in recent years in our laboratory and deals with the subject of soil-bound insecticide residues (LICHTENSTEIN et al. 1977, KATAN et al. 1976, KATAN and LICHTENSTEIN 1977, and FUHREMANN and LICHTENSTEIN 1978).

Closely related to the problem of unextractable, bound residues is that of the persistance or behavior of agricultural chemicals in the environment, in particular in soil.

* Department of Entomology, University of Wisconsin, Madison, Wis. 53706. Presented in part at the April 1979 US-ROC Cooperative Science Program seminar on "Environmental Problems Associated with Pesticide Usage in the Intensive Agricultural System," Taipei, Taiwan, Republic of China, as sponsored by the National Science Foundation (U.S.A.) and the National Science Council (R.O.C.).

For many years depletion curves indicating the persistence or rate of disappearance of insecticides applied to soils have been published. Typical depletion curves were obtained in 1964 by our group after applying insecticides at the same rate, the same time, and by the same methods to loam soil field plots near Madison, WI (Lichtenstein 1975). These depletion curves indicated that compounds such as aldrin or dieldrin persist in soil considerably longer than malathion, methylparathion, or parathion. Frequently the decline in detectable residues has been associated with such terms as "disappearance" or "loss" or "volatilization." However, the apparent disappearance of a pesticide from soil can be due to our inability to detect its residues by conventional procedures. One reason a chemical cannot be detected is that the compound or its degradation products cannot be extracted from soil, thus they are "invisible." Those residues which can be extracted long after their application are the "visible," persistent ones.

The use of radiolabeled pesticides in laboratory studies has made it possible to detect unextractable soil residues. Combustion or strong hydrolysis of extracted soils can release these unextracted or bound ^{14}C-residues. The problem of these bound residues, however, is complicated since present methods for their release or liberation also result in the destruction of their identity. Insight into the mechanism of binding of pesticide residues to soils might shed some light on the nature of the residues and their potential release.

Utilizing ^{14}C-ring labeled parathion, the amount of unextractable or bound ^{14}C-residues in a sandy and a loam soil were determined by combustion to $^{14}CO_2$, after the soils had been extracted 3 times with benzene-acetone-methanol (1:1:1) (Katan et al. 1976). Depletion curves of extractable parathion residues established during a 1-mon incubation period were similar to those established under field conditions and resulted finally in recoveries of 30 to 36% of the insecticide dose applied to the loam soil. With a steady decrease of extractable residues over a 1-mon incubation period, an increase of unextractable, bound ^{14}C-residues occurred. This resulted finally in total recoveries of extracted plus bound residues, which amounted to 80% of the applied radiocarbon. Attempts to exhaustively extract these bound ^{14}C-residues with a variety of solvents ranging in polarity from benzene to water failed to further extract a significant amount. The rate of binding of ^{14}C-residues was highest in the loam soil and was related to the activity of soil microorganisms. In soils, sterilized by gamma irradiation or by autoclaving, the binding of ^{14}C-parathion was reduced by 58 to 84%. Under anaerobic conditions, created by flooding soils with water, the rate of binding of ^{14}C-compounds doubled. The amount of bound residues decreased from 67 to 16% when soils had been sterilized prior to the insecticide treatment and flooding. Reinoculation of this soil with microorganisms fully reinstated the soil binding capacity. Incubation of the moist treated soil under nitrogen also increased the formation of bound residues, while incubation at 6°C rather than at 27°C inhibited binding. When these experiments were repeated using ^{14}C-ethyl rather than ^{14}C-ring labeled parathion, identical results were obtained which indicate that the bound residues apparently contain both the aryl and alkyl portions of the parathion molecule. This information led us to suspect that amino-parathion, which contains both the aryl and alkyl portions of the mole-cule and is formed under anaerobic conditions, might be the bound residue. In addition, earlier data from our laboratory (LICHTENSTEIN and SCHULZ 1964) indicated that while parathion residues could be extracted and detected by thin-layer chromatography in loam soil extracts after several weeks of soil incubation, aminoparathion and p-aminophenol, could not be recovered after a 1-day soil incubation. When this experi-ment was repeated in 1976 using radiolabeled compounds we found that 49% of applied ^{14}C-aminoparathion could not be extracted from the soil 2 hr after its application, while only 1.6% of applied parathion was bound in 2 hr (KATAN and LICHTENSTEIN 1976). Comparison of the binding of all the nitro and amino analogs of parathion during a brief 2-hr incubation with loam soil indicated that in all cases the amino com-pounds were bound to a much greater extent than the nitro compounds.

The role of microorganisms in producing ^{14}C-parathion-derived bound residues and the mechanism of production of soil bound residues was also investigated by incubating ^{14}C-ring-parathion in soil-free culture media that had been inoculated with soil microorganisms (KATAN and LICHTENSTEIN 1976). The amounts of ^{14}C-compounds in culture super-natants, that upon addition to soil became unextractable, increased up to 12 hr of microbial culture incubation, when 43% of the applied radio-carbon was bound after a 2-hr soil incubation period. The increase in soil bound residues was correlated with a decrease in the amount of parathion in the microbial culture and a concomitant increase in the appearance of the major degradation product, aminoparathion.

These data indicate that the production of soil-bound residues of parathion occurs in 2 steps. First microorganisms convert parathion to aminoparathion and then this aminoparathion becomes rapidly bound to soil. These soil bound residues are unextractable and therefore not detected in routine residue analyses.

Experiments conducted with ^{14}C-phorate (LICHTENSTEIN et al. 1978) indicated that 26.4% of the applied residues had become soil-bound within 1 wk of incubation. Contrary to results with ^{14}C-parathion, bind-ing of ^{14}C-phorate residues did not increase after 1 wk. Further investi-gations (LICHTENSTEIN et al. 1977) pertaining to the extractability and formation of bound ^{14}C-residues in an agricultural loam soil were con-ducted with the "non-persistent" insecticides ^{14}C-methyl parathion and ^{14}C-fonofos (Dyfonate®) and with the "persistent" insecticides ^{14}C-dieldrin and ^{14}C-p,p'-DDT. With ^{14}C-methyl parathion only 7% of the applied radiocarbon was extractable 28 days after soil treatment, while ^{14}C-bound residues amounted to 43% of the applied dose.

Field studies conducted in 1968 and 1969 indicated that fonofos has a half-life in Plano silt loam soil under Wisconsin summer conditions of about 28 days (SCHULZ and LICHTENSTEIN 1971). In later studies, this loam soil was treated in the laboratory with ^{14}C-ring- or ^{14}C-ethyl-labeled fonofos and incubated for various periods. After 28 days about 47% of the radiocarbon was extractable and most of this was fonofos. However, unextractable ^{14}C-residues increased with incubation time resulting after 4 wk in 35% of the applied residues being soil-bound. These residues were of course not detected in the field study. Results using ^{14}C-ring- or ^{14}C-ethyl-labeled fonofos were very similar indicating that the bound residues probably do not involve a cleavage product. Contrary to results obtained with parathion the binding of fonofos does not appear to be dependent on microbial activity. While irradiation or autoclaving soil prior to insecticide treatment and incubation significantly reduced the binding of parathion and flooding enhanced it, fonofos binding was not reduced by irradiation. Autoclaving reduced binding somewhat, possibly due to an alteration of soil structure, and flooding slightly reduced fonofos binding. Smaller amounts of soil-bound residues had been formed with the "persistent" insecticides amounting after 28 days to only 6.5% of the applied ^{14}C-dieldrin and to 25% of the applied ^{14}C-p,p'-DDT, while 95 and 72%, respectively, were still recovered by organic solvent extraction. They differed from the organophosphorus compounds in their relatively low binding properties and their high extractability from soils.

The question of the potential biological availability of bound insecticide residues was investigated (LICHTENSTEIN et al. 1977) by testing the insecticidal activity of bound residues from ^{14}C-fonofos and ^{14}C-methyl parathion-treated soils with fruit flies (Drosophila melanogaster Meigen). With soils containing unextractable radiocarbon at the insecticide equivalent of 3 ppm, no mortalities were observed during a 24-hr exposure period to the soil and only slight mortalities occurred during an additional 48-hr exposure period. However, with soils to which the insects were exposed immediately following the insecticide application at the same concentration as the unextractable radiocarbon (3 ppm), 50% of the flies had died within 2 to 3 hr after fonofos application and within 18 to 20 hr after soil treatment with methyl parathion. It appears, therefore, that bound insecticide residues are not only unextractable, but they are also less active biologically.

Experiments were also conducted to study the release and availability of unextractable, soil-bound residues of ^{14}C-ring-methyl parathion and the potential pick up of these ^{14}C-residues by earthworms and oat plants (FUHREMANN and LICHTENSTEIN 1978). Data from this investigation indicate that unextractable soil-bound insecticide residues are not entirely excluded from environmental interaction. After incubation of soil treated with ^{14}C-methyl parathion for 14 days, and exhaustive solvent extractions, bound residues remaining in this soil amounted to 32.5% of the applied insecticide. However, after worms had lived for 2 to 6 wk in this previously extracted soil containing only bound residues or several crops of oats had grown in it, sizable amounts of ^{14}C-residues were found in these organisms. Earthworms which lived in the soil for 6 wk contained a total of 2.7% of the ^{14}C-residues which could not be extracted from these soils, while 3 crops of oat plants, each grown for 2 wk, contained a total of 5.1%. The majority of previously soil-bound ^{14}C-residues taken up by earthworms (58 to 66%) again became bound within these worms, while most (82 to 95%) of the ^{14}C-residues in oat plants were extractable. Greens of oat plants contained 46 to 62% of the ^{14}C-residues recovered from plants. Most of the ^{14}C-residues in oat greens were benzene-soluble while most of the ^{14}C-residues in the seeds and roots were water-soluble.

Because soil-bound insecticide residues can be released from soil by these organisms, any loss in toxicity due to binding should not be regarded as permanent. Even if release of nontoxic compounds occurs, interaction with other chemicals in the environment cannot be disregarded. The release and potential biological activity of these bound residues certainly warrants further study. In view of the above finding, the expression "disappearance" and "persistence" of pesticides, so widely used during the last 2 decades, should be reassessed to consider the bound products.

III. Transfer of soil residues into crops

To compare the soil persistence and metabolism, plant uptake, and metabolism of 6 different insecticides in 2 different soil types under controlled environmental conditions, a study was conducted using the facilities of the University of Wisconsin Biotron (FUHREMANN and LICHTENSTEIN 1979). The 6 insecticides represented organochlorine (^{14}C-DDT, ^{14}C-lindane), organophosphorus (^{14}C-fonofos, ^{14}C-parathion, ^{14}C-phorate), and carbamate (^{14}C-carbofuran) compounds, with water solubilities ranging from 0.001 to 320 ppm. This range was felt to be important because water solubility affects insecticide mobility in soils. Their environmental fate was compared in loam and sandy soils and in oat plants grown in the insecticide-treated soils under controlled laboratory conditions. The total amounts of [4]C-residues recovered from loam soils plus oats were similar with DDT and carbofuran and were higher than those observed with the other insecticides. However, most of the ^{14}C-DDT residues remained in the soils, while most of the ^{14}C-carbofuran residues were recovered from the plants in the form of carbofuran and 3-hydroxycarbofuran. ^{14}C-Residues of all insecticides were more persistent in the loam than in the sandy soil. Sand-grown oats consistently took up more ^{14}C-insecticide residues than loam-grown oats. The more water-soluble insecticides ^{14}C-phorate and ^{14}C-carbofuran were more mobile and also were metabolized to a greater extent than the insecticides of lower water solubility. Unextractable, bound loam soil residues ranged from 2.8 to 29.1% of the applied doses of ^{14}C-DDT, and ^{14}C-parathion, respectively. Bound ^{14}C-residues were lower in the sandy soil than in the loam soil; however, plant-bound ^{14}C-residues were higher in oats grown in the sandy soil than in loam-grown oats. Insecticide metabolites recovered from soils and plants were identified and quantitated whenever possible. The oxygen analog metabolites of the organophosphorus insecticides were most abundant in the sandy soil and the oats grown therein.

The comparative data presented illustrate the importance of 3 factors for determining the environmental fate of insecticides: the insecticide itself, its water solubility, and the type of soil to which it is applied. The chemical nature of the insecticide determines its susceptability to degradation processes, its affinity for soils, its volatility, and its water solubility. The compounds with greater water solubility are more mobile and are taken up by plants to a greater extent. They also appear to be more susceptible to degradation. Insecticide residues are more available for volatility, plant uptake, and degradation in soils with low organic matter (sandy) as compared to more absorbent soils (loam). It is apparent that no single factor can be used to predict the environmental fate of insecticides. Only knowledge of the interaction of several factors will provide this information.

Summary

Since ^{14}C-labeled insecticides have become available, formerly unextractable, bound residues can be detected by combustion of the ^{14}C-containing material to ^{14}CO$_2$. In this way we have become aware of the existence of formerly nondetectable residues. Although the nature of these bound residues is in most cases unknown, our concept of "persistent" and "nonpersistent" pesticides might have to be reconsidered. With NO$_2$-group-containing compounds such as parathion and methyl parathion, it was found that soil microorganisms reduce these insecticides in soil to amino-derivatives. These, in turn, become bound to the soil and are considered unextractable. Thus, a steady decrease of extractable ^{14}C-parathion residues in soils over a 1-mon incubation period was accompanied by an increase of unextractable, bound ^{14}C-labeled residues, resulting finally in total recoveries of extracted plus bound residues of 80 to 87% of the applied radiocarbon. Soils containing bound residues were nontoxic to fruit flies. Soil sterilization resulted in a reduction of binding by 58 to 84%. Under flooded (anaerobic) conditions, the binding of compounds labeled with ^{14}C doubled, and parathion was reduced to aminoparathion. Reinoculation of sterilized flooded soil fully reinstated the binding capacity. ^{14}C-Aminoparathion was preferentially bound to soil, since its binding within 2 hr was 30 times greater than that of ^{14}C-parathion. Although ^{14}C-residues cannot be extracted from soils by conventional means, earthworms and oats were able to release them from the soil, and also metabolize them within their tissues. After incubation of soil treated with ^{14}C-methyl parathion, and exhaustive solvent extractions, unextractable-bound residues in this soil amounted to 32.5% of the applied insecticide. However, after worms had lived for 2 to 6 wk in this previously extracted soil or several crops of oats had grown in it, sizable amounts of ^{14}C-residues were found in these organisms. Once they had penetrated into the animal or plant tissue, they were translocated and found partially in an unextractable, bound form or as benzene- and water-soluble ^{14}C-compounds.

To study transfer of soil residues to crops, a sandy soil and a silt loam were treated with one of 6 insecticides of different water solubilities (0.001 ppm to 320 ppm) and plants were grown in the soil. The amounts of ^{14}C-compounds which penetrated into the plant tissue were dependent upon the water solubility of the insecticide and the soil type, i.e., most of the ^{14}C-compounds were picked up from a sandy soil which had been treated with the insecticide exhibiting the highest water solubility. The total amount of ^{14}C recovered from soils plus plants was similar with DDT (water solubility 0.001 ppm) and carbofuran (water solubility 320 ppm). However, with DDT most of the insecticide remained in the soil while with carbofuran most of the recovered insecticide residues plus metabolites were associated with the greens. The amounts of ^{14}C bound to plant tissue as well as the amounts of detoxification products in plant tissue increased with increasing water solubilities of the insecticides.

References

FUHREMANN, T. W., and E. P. LICHTENSTEIN: Release of unextractable soil bound ^{14}C-methylparathion residues and their uptake by earthworms and oat plants. J. Agr. Food Chem. 26, 605 (1978).
—— —— A comparative study of the persistence, movement and metabolism of six ^{14}C-insecticides in soils and plants. J. Agr. Food Chem. 28, 446 (1980).
KATAN, Y., and E. P. LICHTENSTEIN: Mechanism of production of soil bound residues of ^{14}C-parathion by microorganisms. J. Agr. Food Chem. 25, 1404 (1977).
——, T. W. FUHREMANN, and E. P. LICHTENSTEIN: Binding of ^{14}C-parathion in soil: A reassessment of pesticide persistence. Science 193, 891 (1976).
LICHTENSTEIN, E. P., and K. R. SCHULZ: The effects of moisture and microorganisms on the persistence and metabolism of some organophosphorus insecticides in soils, with special emphasis on parathion. J. Econ. Entomol. 57, 618 (1964).
——, KATAN, Y., and B. A. ANDEREGG: Binding of "persistent" and "nonpersistent" ^{14}C insecticides in agricultural soil. J. Agr. Food Chem. 25, 43 (1977).
——, T. T. LIANG, and T. W. FUHREMANN: A compartmentalized microcosm for studying the fate of chemicals in the environment. J. Agr. Food Chem. 26, 948 (1978).
SCHULZ, K. R., and E. P. LICHTENSTEIN: Persistence and movement of Dyfonate in field soils. J. Econ. Entomol. 64, 283 (1971).

Manuscript received May 12, 1979; accepted January 24, 1980.

Interpreting pesticide residue data at the analytical level

By

Francis A. Gunther[*]

Contents

I. Introduction

The natures, locations, and quantities of pesticide residues in food-stuffs are important in the realm of the public health; these residues in animal feeds are important, also, in animal husbandry, for they conceivably could affect the health and well-being of the animal as well as of the ultimate consuming public through possibly pesticide-contaminated edible animal products. Properly used under the now legalized aegis in many countries of "good agricultural practice," with all the constraints attached thereto, there should be no nutritional or other health-related adverse effects from pesticide or pesticide-derived residues persisting in the agricultural environment. It is the occasional misuse or illegal use

[*] Department of Entomology, University of California, Riverside, CA 92521. Presented in part at the April 1979 US-ROC Cooperative Science Program seminar on "Environmental Problems Associated with Pesticide Usage in the Intensive Agricultural System," Taipei, Taiwan, Republic of China, as sponsored by the National Science Foundation (U.S.A.) and the National Science Council (R.O.C.).

of pesticide chemicals,[1] however, which could result in possibly deleterious residues persisting from attempted pest-control applications into human foodstuffs or animal feeds. It is not the intent here to dwell on the many complications that can result from over-contamination with pesticide chemicals of the nonedible agricultural environment such as soil, runoff and other initially agricultural waters, nontarget plants, wild animal and aquatic life, and air.

No matter how carefully prepared and detailed are the government regulations specifying pesticide, formulation, dosage, method of application, timing, crop, pest, and harvest-interval, there will be misuses and occasional illegal uses of pesticide chemicals in widespread agricultural practice from ignorance, economic necessity, or haste. The early detection of these abuses therefore becomes of major public health importance, early enough to prevent the pesticide-contaminated item from reaching the market or even from being consumed on-site.

II. Scope of problem

This reliable detection of natures and quantities of aberrant pesticide residues in foodstuffs and feeds is an analytical matter of concern to us all. It is also abundantly clear that definitive residue analytical data are valueless without some reliable, efficient, and effective system of reporting them to government or other officials who can make incisive decisions to condemn the offending item and then promptly see that it is removed from channels of trade. This analytical effort requires:

1. Organization within a country for food control of domestically produced agricultural products as well as counterparts world-wide for food control of both exports and imports. This organization involves technical (residue analytical chemists), administrative, legislative, and punitive arms with means for rapid intercommunication and action.

2. A research arm qualified and competent to develop reproducible residue analytical methodology adequate for the necessary food control of that country. This requirement embraces not only domestic production and consumption but also imported and exported foodstuffs and feeds; not only must the country in question meet its own food-quality criteria but also must it analytically monitor exported foodstuffs and feeds to assure compliance with the residue quality criteria of a foreign market and probably even occasionally monitor imports to assure continuing conformity with its own domestic requirements. These activities necessitate a research arm of considerable scope, versatility, and knowledge of pest-control chemicals in current use essentially around the world in these days of increasing gourmet appetites among consuming publics everywhere.

---

[1] This term as here used includes congeners in the formulation used as well as *in situ* and other alteration products significantly toxic to warm-blooded animals or capable of potentiating any significantly toxic chemical species present.

3. A monitoring arm expertly staffed and well enough equipped to cope efficiently with large numbers of diverse samples conceivably containing diverse chemicals of public health concern. Depending upon local circumstances, this monitoring arm may need mobile residue-analytical components to follow harvests most expeditiously, as exist in California and Florida, U.S.A. If it has fixed sites there must be means of getting field samples to the laboratory quickly and without residue deterioration. In turn, the laboratory must analyze the samples without delay so as to prevent an illegally contaminated product from reaching the market.

4. These monitoring laboratories must be prepared routinely to characterize the chemicals in question as well as to measure their amounts with meaningful reliability. Such identification at normal residue levels involves a preponderance of evidence and is never deduced from a single analytical technique or instrument. In turn, measurement data incorporate a great deal of conservative discretion on the part of the knowledgeable residue analyst in establishing the real residue value divulged by the technique(s) utilized, from application to sampling to processing to cleanup to determination ("estimation" at most currently extant residue levels in foodstuffs and feeds).

5. To have any value in the present sense, the residue data so obtained must be communicated quickly and in incisive detail to those officials empowered to make decisions when a commodity contains residues exceeding the local or other tolerance limits. If a decision is unfavorable, it must be acted upon quickly to keep the offending commodity out of the hands of the consumer. This action implies the existence of still another mechanism to halt—for this particular commodity—the often fast-moving chain of events: farmer → wholesaler → distributor → retailer → consumer. Frequently this entire chain of events requires only 12 to 24 hr for highly perishable commodities, such as strawberries. As an epilogue to this condemnation action process, there must, of course, be carefully thought-out means for the irretrievable disposition of condemned foodstuffs and feeds.

6. Finally, and properly beyond the scope of the present report, there must be within any country using pesticides in agriculture an efficient and current mechanism for keeping the farmer (grower) informed of penalties consequent to violating "good agricultural practice." As a corollary to this essential need, there should be some kind of governmental organization to help prevent—and hopefully to detect in the field—violations of "good agricultural practice," whether they be caused by ignorance, economic pressures, carelessness, indifference, or haste.

III. Food control organization

Governmental regulations and actions to maintain and preserve high quality in the world's food supplies have long been termed "food control."

This term also embraces control of the natures and amounts of residues of parent pesticide chemicals and sometimes their toxic *in situ* alteration products, if present. Various government have used various means to assure this integrity of foodstuffs and feeds with regard to permissible pesticide residue contents, but to date the most technically acceptable means have involved the tolerance concept and/or the well-established concept of minimum interval to harvest. The former concept requires residue analytical facilities and personnel, as well as an efficient organization to insure that analytical data are representative of production or import and that over-tolerance commodities will not reach the consumer. The latter concept, based upon residue degradation and persistence curves (GUNTHER and BLINN 1955), assures that under normal field conditions in the producing country the initial deposit will degrade or otherwise attenuate with time until it reaches a safe level.

Importantly for some countries, the minimum interval concept does not require monitoring analytical facilities, which *in toto* are elaborate, expensive, and complex in overall organization to achieve efficiency of purpose. Reliance upon this concept, however, does require the analytical establishment by someone that the minimum intervals adopted are indeed adequate for that pesticide, that formulation, that dosage, that method of application, that time of year, and that crop grown in that country. In general, minimum intervals are selected to err on the safe side so they are broadly applicable from one country to another, but confirmation of this generality is reassuring in every instance of application.

Double assurance, selected by many countries, is achieved by adopting both concepts so as to uncover the occasional aberrant sample of foodstuff or feed which invariably—even under "good agricultural practice"—would seem to result from misuse of the pesticide at time of application.

To be at all effective, there must be means not only to detect violations of both concepts but also to remove the resulting "illegal" commodities from the market as well as to punish the original offenders (farmers, advisors, pesticide applicators) to assure nonrepetition of the incidents.

IV. Food control research arm

Somewhere within most agriculturally producing and/or consuming countries there must exist pesticide residue research arms capable of developing or adapting residue analytical methodology to the particular needs or circumstances of that country. Many food control administrators feel that major toxic metabolites or other *in situ* pesticide alteration products should be monitored or otherwise controlled along with the parent chemicals. If this exaggerated precautionary policy is adopted, the research and monitoring tasks both become unmanageable. In those countries where the administrators have this inclination, an achievable compromise has sometimes been made—the research arm establishes natures and proportions of these alteration products in field-aged residues for general guidance in helping interpret harvest-time residue data involving only the parent compound as analyte. In other words, extreme conservatism is called for in the interpretation of residue data where possibly significant amounts of toxic alteration or other products are likely to be present.

It is obvious that the development of a new pesticide residue analytical method, with a given crop as substrate, is a research problem. It is not always so obvious that the adaptation of an existing residue analytical method to a new crop or other food or feed commodity is also a research problem, often requiring weeks of skilled effort. Pesticide residue detection and determination is still an art, and especially not a simple science to be entrusted to untrained, inexperienced personnel. Unwarranted assumptions by laboratory leaders and other administrators are usually made of the simplicity of the detection and measurement and of the reliability of the method; such difficulties make the measurement of all but the simplest parameters, and the interpretation of the results, by other than qualified [residue] analytical chemists a dubious procedure (*American Chemical Society* 1978).

Although most current pesticide residue analytical methods are gas chromatographic for final cleanup and measurement, the "old-fashioned" largely colorimetric methods are usually still appropriate, especially where maintenance (servicing) of complex electronic equipment is a problem. For example, most pesticide residue tolerances around the world range from a few hundredths of a ppm to 1.0 or more ppm. This range is easily achievable with many colorimetric methods. Similarly, thin-layer chromatographic methods for segregation of sought analyte followed by colorimetric assay are also still appropriate. It is only when residues may be present at ppb or ppt levels (e.g., water) that colorimetric methods require prohibitively large samples of substrate. It is just as easy to process 500 g of substrate as it is to process 1 to 10 g, and much more analytically reliable (see later); similarly, the cleanup of the total extractives from a 500-g subsample is not more difficult than the cleanup of the total extractives from a 10-g subsample.

Despite this attractive and possible eventual monitoring use of colorimetric or other "simple" residue analytical methods, the residue research arm should be equipped with UV and IR spectrophotometers, glc, and glc/ms, because each step of a new or revised residue analytical method must be confirmed as to specificity, reliability, and efficiency before the method is put into routine use, whatever the end detection and measurement system: the residue analyst must have assurances that what he is measuring is indeed the sought species. Zweig (1978) has very capably recently reviewed modern instrumental trends in pesticide residue analytical detection and determination. The technical literature on pesticide residues in foodstuffs, soil, and water now contains many examples of earlier mistaken analytical identities—such mistakes can allow a potentially harmful residue to reach the consumer or they can prevent an otherwise useful pesticide chemical from commercial utilization. Notorious examples of mistaken identities from too casual glc analytical efforts include sulfur interpreted as aldrin, PCBs interpreted as DDT, p-dichlorobenzophenone interpreted as dicofol, unknown artifacts or sloppy techniques interpreted as DDT in rainwater over London and in Arctic snow, inorganic phosphates in human urine interpreted as alkyl phosphates from presumed exposure to OP compounds, numerous claims of ppb and even ppt levels of many pesticides in various waters and crops (see later), the interpretation by a regulatory agency of o,p'-dicofol as heptachlor, and the demonstration of several organochlorine insecticides in several soils sampled and sealed in 1910, 30 years before the organochlorine insecticides were commercially available [see GUNTHER (1971) for other examples, particularly those involving quantitation]. A recent collaborative study arranged by the author (unpublished) involved 24 pesticides added in 0.1 to 10 ppm amounts to virgin oil of orange: 4 reputable residue analytical laboratories missed as many as 8 chemicals, reported 4 that were not present, and did equally poorly on quantitation for those chemicals actually present (3 to 140% deviation from the true values with the greatest deviation at the 0.1-ppm levels).

V. Food control monitoring arm

The monitoring arm of any pesticide residue regulating organization is the key operation, for it is intended to establish which commodities in hand are safe to eat and which are not safe to eat. This operation must be efficient, reliable, and prompt because often there is not much lag time between harvest and marketing. The passage of an excessively contaminated lot or shipment could conceivably cause harm, whereas the overly long retention of a truly "clean" lot or shipment will result in economic hardship to the producer.

For these reasons this arm—or, with prompt cooperation, the research arm—should be equipped to confirm the probable identity and also, by another method, the amount of analyte found in any suspected overtolerance sample. Mistakes in this arena frequently result in lawsuits.

In countries where there is truly intensive agriculture it may be expedient to have mobile monitoring laboratories to follow a geographically progressing harvest, as earlier mentioned, for, again, interval between sampling and clearance or condemnation is usually economically critical.

In the usual agricultural situation, a given monitoring laboratory will have to deal with only a few pesticides and/or a few commodities from that particular region. If samples are transported from a large growing area to a central laboratory, however, that laboratory must be equipped and staffed to handle large numbers of diverse substrates containing diverse chemicals, often of unknown nature so that so-called "multiple residue methods" should be available and ideally supplemented by rapid bioassay procedures for prompt "aye" or "nay" clearance as to probable toxic hazard, whatever the natures of the persisting residues.

Confronted by the thus confirmed residue data for an offending sample, the monitoring analyst-in-charge must decide the reliability of the claimed identity of analyte and the numerical significance of the claimed amount, then communicate this information quickly to those officials responsible for authorizing and executing commodity seizures or other immediate stop-sale measures. It is hardly necessary to reemphasize that the analyst-in-charge must assure himself there can be little question that the claimed analyte is indeed present in the claimed realistic amount—a commodity seizure is serious business.

Also needless to dwell upon is the essentiality in any monitoring program to assure the integrity of any sample accepted by the monitoring laboratory; "integrity" here means minimum time between sampling and delivery to the laboratory, precautions to insure sample labels are correct, cold or preferably frozen transport of sample to minimize analyte deterioration, and prompt analytical attention to the sample at the laboratory end.

It is beyond the scope of this report to discuss pesticide residue sampling philosophy and procedures; suffice it to record that in any pesticide residue investigation, by administrative edict, it must be made clear whether samples are to be residue-representative or residue-maximizing. The analyst-in-charge should then be involved in establishing sampling procedures, hopefully to be authenticated by the research arm above; authenticating or verifying a sampling procedure is neither a simple nor an easy matter, viz., a paddy of rice vs. a field of watermelons vs. a patch of strawberries vs. a truckload of lettuce.

VI. Interpretation of residue data

Pesticide residue regulation administration must be presented with incisive residue data, for incisive decisions have to be made. The difficulties inherent in pesticide residue investigations make the measurement of all but the simplest parameters, and the interpretation of the results, by other than qualified residue analytical chemists a dubious procedure; the result has been masses of residue data [in the literature] that frequently are useless for the intended purpose (*American Chemical Society* 1978). I have often stated in public that most pesticide residue data for crops and crop products, as single items, below 0.1 ppm are analytically not very meaningful; adequately replicated applications, samples, and sampling sites can lend credence to such data within broad ranges (GUNTHER 1969, 1970, and 1971):

Residue	Latitude
10 ppm	±10%
1 ppm	±10%
0.1 ppm	±20%
0.01 ppm	±50%
0.0001 ppm	±200% (usually more)

Frehse and Timme (1978) have expanded this latitude concept into a more technical format, presenting the mathematical and graphical principles of the concept and expositing its utility for settling pesticide residue analytical controversies in relation to maximum residue limits [tolerances]. Their conclusions agree essentially with the above simplistic guidelines. Actual numbers from published examples have been reported in the technical literature (e.g., Gunther 1971). The important conclusion is that reported pesticide residue values in crops and other plant and animal parts "may vary considerably from the 'true' values" (Corneliussen 1970).

The flamboyant adoption of the gas chromatograph as the only [unsupplemented] pesticide residue analytical tool has resulted in this instrumental technique being strained to the utmost, oftentimes even in experienced hands. A properly operated gas chromatograph is a remarkable cleanup and detection device, with properly selected detectors, and under ideal conditions in the absence of complex substrate extractives. It is not an identifying device, under any circumstances, although its proper use in experienced and knowledgeable hands may support identification of a particular analyte.

a) Identity—preponderance of evidence

At the usually low pesticide residue levels encountered world-wide with foodstuffs and feeds, and their products, it is hardly possible unequivocally and firmly to identify the persisting residue as a single, specific compound. These levels generally range from a high of about 10 ppm in "good agricultural practice" and about 30 days after application to several hundredths of a ppm, or a 3-decade range of concentrations. At the 1-ppm level there will be a μg of analyte/g of substrate which, in turn, often contains 10% or more by weight of total organic solvent extractables. This extraction operation, then, affords somewhat less than 1 μg of analyte plus at least 100,000 μg of other organic chemicals in complex admixture. Adequately and reproducibly separating the sought chemical from these congeners is known as "cleanup" and under the above circumstances it can only approach high efficiency. The final cleaned up mixture to be scrutinized by the analytical "detector" often contains analyte mimics to which the detector will respond, no matter how carefully selected and adjusted. Except for a mass spectrometer, all presently available analytical detectors respond to only selected portions of the molecules to which they are exposed; since they are "aware" of only the sought portions, not the entire sought analyte, such detectors will lie to the analyst where they "see" the moiety in question in any foreign molecule in the analytical stream. The pesticide residue analyst must therefore be constantly on the alert to detect and suitably to compensate for such false reports, both positive and negative.

Similarly, a gas chromatograph is basically a separation or cleanup device and identification-reliance upon raw retention times or retention volumes is hazardous and foolish. The technical literature contains abundant examples of later-proven mistaken identifications and consequent false indictments of major useful pesticide chemicals, as discussed earlier.

Similar constraints apply to other cleanup and detecting techniques at the residue levels encountered in modern pest-control practices. As pointed out earlier, the concentration of the sought analyte in the total extractives mixture is very low (ca. 1:100,000) and the extractive mixture is chemically complex, with abundant opportunity for functional group duplication with the analyte as well as functional group mimicry. Cleanup techniques utilize principally the hopefully unique specific chemical and/or physical properties of the functional groups of the analyte vs. those of the coextractives. Thus, normally, gross separations (e.g., solvent partitionings, hydrolysis, oxidation, steam distillation, etc.) to reduce bulk of coextractives are followed by refined segregative techniques such as chromatography (gel, HPLC, TLC, glc, paper, column), countercurrent distribution, electrophoresis, and others to achieve final and hopefully adequate separation of the analyte from almost everything else. HPLC and glc are superb techniques for approaching the final goal, if properly carried out with properly functioning equipment.

As pointed out earlier, however, proper chromatographic peaks at the proper time do not necessarily identify—they *support* identification. Unequivocal identification at the levels encountered in practice is hardly possible, even with a mass spectrometer. Claimed identifications can be buttressed at these levels, however, even without a mass spectrometer. For example, with usual laboratory equipment and techniques the proper use of internal standards, co-chromatography, glc vs. TLC (sometimes) (ELGAR 1971), Beroza p-values (BEROZA et al. 1969), multiple detectors (especially those that respond to different functional groups or other unique moieties), drastically different cleanup procedures, etc., can contribute significantly to a preponderance of evidence that makes highly probable the to-be-claimed identification. Double gas-liquid chromatography on different columns will not necessarily support identification (see ELGAR 1971).

b) Real numbers

At the levels encountered in good agricultural practice the decimal significance of fractional ppm of persisting residues should be doubted by pesticide residue analysts everywhere. These numbers will be used by someone—usually a nonanalytical chemist—with the administrative responsibility of having to make a decision as to whether a particular residue is safe or not safe, and a decision of unsafe may result in the required destruction of an entire shipment of a food or feed commodity. Furthermore, this administrator, not knowing all the details and uncertainties in pesticide residue analytical methodology, will be inclined to assume the difference between, for example, 0.125 and 0.126 ppm is real, when in actuality both these numbers really mean 0.1 ppm ± 20% (see earlier) for field samples of soil, produce, or animal product (not necessarily air or water). Misinterpretations of this sort can have drastic and serious repercussions as discussed earlier.

People who are not trained as analytical chemists are inclined to regard analytical chemistry as an exact (precision) science, producing nothing but real numbers no matter how many decimals are carried out.

The ideal analytical method responds only to a single substance in the mixture presented to it and it reveals the exact amount of that substance present in the parent substrate. In actuality, analytical methods deviate from this ideal performance in characteristics that include specificity, selectivity, sensitivity, and lower limit of detection (American Chemical Society 1978).

In pesticide residue chemistry, specificity refers to the number of substances to which a method will respond: today there are no truly specific pesticide residue analytical methods. Selectivity refers to the preferential response to any one of several related substances in a mixture of foreign substances or interferences. Sensitivity properly is the slope of the standard curve, or the smallest incremental amount of analyte to which the method will reliably respond. Lower limit of detection (detectability limit) is the smallest absolute amount to which the method will reproducibly respond. Accommodation of instrumental and other "noise" may make the lower limit of detection somewhat greater than the sensitivity; thus, for example, the sensitivity may be 0.1 ppm, whereas the "noise" may also be 0.1 ppm, in which case different pesticide residue analysts have felt the minimum amount reproducibly detectable is 0.15 ppm (FREHSE 1964), 0.2 ppm (GUNTHER, unpublished data), or even 0.4 ppm (SUTHERLAND 1965). Furthermore, it should be emphasized that there is no analytical method which will demonstrate the complete absence of any analyte from any sample, but only that it is "not detectable." It should be noted here that both sensitivity and minimum detectability under ideal conditions can be markedly different when substrate extractives are present. It is also appropriate here to decry the too-common reporting of "trace" amounts of residues. "Trace" means a very small quantity not determined because of minuteness. Even when defined numerically in each usage, as, for example, "trace means less than 0.1 ppm," this habit should be discouraged. To the uninitiated, "trace" implies that some of compound x is present. On the other hand, the analyst used "trace" because he was uncertain as to identity and as to amount; in other words, the instrumentation indicated a possible but not measurably definite response. Because of this connotation, it is much more acceptable to use "none detectable" (ND) and in each report to define the lower limit of reliable detection in semi-quantitative terms as the total method was utilized with suitably fortified control samples.

The results obtained with any analytical method tend to be sensitive to the skills and to the environment of the analyst (*American Chemical Society* 1978). Thus, the precision, accuracy, and reliability become important as established usually by collaborative tests. *Precision* describes the scatter among the results, irrespective of the true composition of the sample, and may be considered the % relative error in an analysis. *Accuracy* represents the difference between the average result and the true value. *Reliability* refers to the over-all reproducibility of a measurement. Accuracy in total-method pesticide residue quantitation is not easy to achieve, for without the use of radiotracers there is no easy way to establish the efficiency of each analytical step, including the final determination.

The usual pesticide residue analytical scheme, from field sample to final calculation and interpretation, involves the following combination of steps: sampling, sample storage, sample subdivision, extraction (processing), concentration, cleanup, (concentration), determination, calculation, and interpretation.

1. Sampling.—Pesticide applications are not uniform in a field or grove no matter how carefully carried out. Initial deposits, and resulting residues, will vary by as much as 400% from plant to plant, from top to bottom of a plant, from north to south sides, and from outside to inside foliage or fruit (Gunther and Blinn 1955). If sampling methods are to seek maximum residues present, the sampler must know where on a plant residues will be maximum; if they are to be representative, the sampler must include both maximum and minimum specimens.

Long experience by any workers with many substrates has demonstrated that statistically selected samples, by experienced samplers, will show up to a 25% variation in apparent residues present among replicates. Soil samples normally show even greater variations.

This process represents the first big variable affecting the accuracy and reliability of pesticide residue data.

2. Sample storage.—This heading includes sample transport to the laboratory.

As firmly established by Kawar et al. (1973), even frozen storage of some pesticide residue-bearing plant parts will not prevent the time-dependent deterioration of some pesticide chemicals and their alteration products; other workers have occasionally mentioned similar results. In any event, it should never be assumed without proof that frozen storage will maintain the pesticide residue integrity of a plant, soil, or water sample. Likewise, samples should be transported from the sampling site to the laboratory at least iced and preferably frozen, and frozen storage times should be minimal unless it has been clearly demonstrated frozen storage does halt *in situ* residue alteration processes. As a general rule of thumb, frozen storage of more than 30 days requires proof of residue stability, but even this concession is clearly not valid for some materials (KAWAR et al. 1973).

Neglect in this area can cause large and variable losses of residues.

3. Sample subdivision.—This topic has been thoroughly discussed many times (e.g., GUNTHER and BLINN 1955). Suffice it to repeat here that uniformity (homogeneity) of subsamples and often efficiency of extraction (processing) are dependent upon particle size and thoroughness of mixing. After suitable subdivision, subsampling should be done by quartering.

Casual techniques in this area can also result in greatly variable residue analytical values among replicated subsamples.

4. Extraction (processing).—The literature on this subject is voluminous, for each pesticide residue analyst has his favorite organic extracting solvent and extracting procedure, usually dictated by costs of solvents, availability of equipment, and adequacy of ventilation of the processing facility.

The most important factors which can affect the efficiency of extractions of samples of plant and animal parts and of soil, and thus the accuracy and the reliability of the final residue values, are summarized below:

a. Soxhlet extraction is in general to be avoided because of the thermal lability of most pesticide chemicals, particularly in the presence of water (from the sample itself) and of solvent-insoluble ligand generators (also from the sample itself); photodegradation is also a possibility with a few pesticide chemicals during prolonged Soxhlet extractions. Another factor to worry about here is the possibility of escape of pesticide or pesticide alteration product through the Soxhlet chimney through simple volatilization or codistillation.

In addition, Soxhlet extraction with a water-immiscible solvent (except diethyl ether) of a largely aqueous substrate (most plant parts) is not an extraction-efficient operation.

b. Equilibration of the finely divided substrate with a mixture of solvents (see below) is more time efficient and for the present purposes just as satisfactory as exhaustive extraction (GUNTHER and BLINN 1955). In this procedure the solvent-marc mixture is equilibrated with agitation at room temperature, the solvent mixture is decanted, and the equilibration is repeated with eventual combination of decantates. The efficiency of this process in any given laboratory is easy to approximate (GUNTHER and BLINN 1955); so long as conditions are selected to assure reproducibility, efficiency-correction factors are used in the final calculations of residue load present in the substrate.

c. Largely aqueous substrates, finely divided, are best extracted by first mixing with an equal volume of a lower alcohol or acetone then adding 2 or 3 volumes of the water-immiscible solvent and equilibrating the mixture. The final solvent mixture, freed of marc, is watered down to yield the analyte(s) in the immiscible organic phase; if the analyte is appreciably water-soluble it may be necessary to saturate the aqueous phase with salt and multiple extract to achieve adequate partition. n-Hexane has usually been the latter solvent of choice and, from cost considerations, isopropyl alcohol is usually the former solvent. All these considerations have been thoroughly discussed in detail by many authors.

The efficiency of this operation is high and reproducible.

5. **Concentration.**—This operation may or may not be necessary at several stages during the cleanup operations. Recoverywise, the safest technique is the Kuderna-Danish technique (GUNTHER and BLINN 1955), because this compact multiple set-up apparatus is used at atmospheric pressure, is self-rinsing as the level of solution recedes, if properly used (pre-wet Snyder column) results in no variable volatilization or codistillation losses, and needs no attention, for a K-D apparatus cannot run dry.

The vacuum rotary evaporators in such common use today are subject to variable volatilization and codistillation losses, even when carefully tended, and represent major monetary investments requiring such valuable bench space if used in multiples.

6. **Cleanup.**—Satisfactory cleanup techniques are limited only by the imagination of the analyst in terms of the chemical and physical characteristics of analyte vs. substrate. However chosen, they must be reproducible and efficient.

7. **Determination.**—Again, the imagination and knowledge of the analyst are the limiting factors here. Determinative techniques may be legion for a particular problem or, conversely, highly restricted. As discussed earlier, the detector must be reproducible and efficient in performance and as selective as required by the problem at hand. Skill and experience are required to know when a detector is lying and elaborate precautions may be required to keep a detector on the path of truth.

In this category lie the greatest pitfalls against verity of final analytical results. The preceding operations can each cause a hopefully nonvariable but partial loss of analyte. The detector can fail completely to respond at all when analyte is indeed present or it can report the presence of analyte when there is none there. Frequent internal standards, blanks, controls (untreated samples), and fortified controls may be required to assure continuing integrity of detector responses.

8. **Calculation and interpretation.**—Calculations are most safely based upon standard curves prepared from controls and fortified controls, put through the entire analytical process, and not just upon standard solutions exposed to the detection technique. Not only may these two types of curves be displaced with regard to each other but also their slopes may be markedly different one from the other.

Pesticide residue data are not easy to interpret as to the numerical significance of a given value or even several replicated values, as, for example, at harvest time. Presented earlier were the latitude ranges to be associated with such individual data. In practice these ranges are not overly generous unless the residue program is statistically designed, including especially the field sampling protocol and schedules. Thus, under ideal circumstances the residue analyst can establish persistence curves and with them sample-to-sample variation to achieve highly meaningful data. In monitoring programs, however, the analyst normally has to deal with harvest-time samples only, and perhaps even single samples are submitted. In all such instances, the analyst should claim only "apparent" or "probable" parathion, for example, at 2.0 ± 0.3 ppm.

Field sampling efficiency incorporates application irregularities as well, as discussed earlier, and seems (Gunther, unpublished data) to average out with citrus fruits at about ±25% with statistically designed sampling techniques (Gunther and Blinn 1955), *i.e.*, 32 fruits/sample, 4 from each of 8 trees, selected by quadrants at chest height, and replicated twice with 2 samplers circling the trees in opposite directions; in this manner, top-to-bottom, inner-outer, and north-south variations are minimized. The final residue result of such a scheme is, of course, an average (representative) sample, not a maximum residue sample.

Subsampling efficiency can easily be optimized with sufficient care and patience.

Processing efficiency can be highly variable but it can be adequately evaluated with sufficiently carefully prepared fortified blanks and fortified controls. Even without the use of radiotracers there are techniques for quite accurately establishing the processing efficiency of extractable analytes, often before and after hydrolysis to rupture "binding" ligands, as with soil samples. If properly carried out, processing is reproducible and thus correctable.

Concentration losses are similarly correctable, if reproducibility is carefully established over the range of concentrations expected.

Cleanup reproducibility and efficiency are again established with blanks, fortified blanks, controls, and fortified controls, with fortifications over the entire range of interest to the analyst and to the regulatory agency. Cleanup must be reproducible; its efficiency is correctable.

The final determination is normally reproducible and adequately efficient, if instrumental and if properly "policed" to assure maintenance of integrity of final read-out.

Acceptable over-all laboratory efficiency, or "recovery," normally varies from one compound to another, or from one substrate to another with the same compound, from an acceptable low of about 60% to a high of about 110%, if reproducible within ±5%. This efficiency can be used as a correction factor in the final calculations.

The net result of all these deliberations is that a final pesticide residue value has considerable uncertainty, not only as to identification but especially as to amount claimed to be present. This uncertainty must be numerically estimated by the residue analyst and communicated to the administrator or other official who will make the safe-unsafe decision. Only the experienced and knowledgeable residue analyst can so interpret the data from his own laboratory; it is obligatory upon him to do so for the important reasons discussed earlier.

VII. Screening of samples

Most analytical detection devices (instruments) or procedures (e.g., colorimetry) only report, or infer, the degree and maintenance of segregation achieved by the total analytical procedure as well as the amplitude of the signal generated. Thus, choice of a detector often depends upon whether the ultimate analytical objective is research or monitoring. It is outside the province of this report to discuss detailed research in pesticide residue methodology. Monitoring, or "screening" (WESTLAKE and GUNTHER 1967), is the present subject.

There are at least 3 types of screening often used in combinations in the present application:

1. Segregative screening—separating above-tolerance from below-tolerance samples, with the acceptable quantitative latitudes discussed earlier and usually with only one sought pesticide chemical in mind.

2. Constituent screening—detecting a variety of sought pesticide chemicals in the sample, with previously established lower limits of detection.

3. Quantitative screening—determining or otherwise adequately establishing the probable amounts of sought pesticide chemicals present in the sample, again with previously established lower limits of detectability but also with previously established reliability and reproducibility.

VIII. Communication of monitoring data

This subject has been discussed at appropriate times throughout this report. It cannot be overemphasized that prompt communication of incisive pesticide residue data to responsible officials is essential if the integrity of the world's supply of foodstuffs and feeds is to be maintained or bettered.

The data communicated must be sufficiently incisive that the decision-making official is aware of the reliability of individual data, and consequently that a single datum is essentially meaningless and could not possibly be defended in court. Confiscation of a field of produce, a lot, or a shipment for bearing illegal pesticide residues will almost invariably result in a lawsuit. Modern pesticide residue monitoring data are not easy to defend—on scientific grounds—in a court of law.

The moral obligation for the pesticide residue analyst to produce as definitive data as possible cannot be controverted.

IX. Conclusion

Pesticide residue value are only estimations, at best. At the levels encountered in practice there cannot be other than some doubts as to actual identity of claimed residue and serious doubts about the amounts present as revealed by the total analytical procedure utilized.

Summary

Only the residue analyst can properly and incisively interpret his own data. Uniformity of field substrate and of application can be highly variable, but adequate sample replication will help define these parameters for the analyst. Factors such as processing efficiency, variability of cleanup techniques utilized, fortification recoveries, instrumental errors, total precision and reliability, and minimum reproducible detectability are adequately understood only by the analyst for each pesticide in each field substrate situation.

The numbers the analyst produces must be rounded off to accommodate these factors. Furthermore, the identities of claimed analytes at the ppm level are based upon technicological circumstantial evidence, often precarious without elaborate verification by ancillary analytical techniques.

Proper public health decisions based upon detailed pesticide residue data can be made only if the analyst is honest in reporting his data to these officials to accommodate all these uncertainties. These officials probably will not understand these analytical details and their significance; data should therefore be reported as, for example, "probable parathion, 2.0 ± 0.3 ppm." Rarely can a second decimal have any significance in modern pesticide residue methodology with random field samples.

References

American Chemical Society: Cleaning our chemical environment: A chemical perspective. Second ed., pp. 41 ff. and 320 ff. (1978).

BEROZA, M., M. N. INSCOE, and M. C. BOWMAN: Distribution of pesticides in immiscible binary solvent systems for cleanup and identification and its application in the extraction of pesticides from milk. Residue Reviews **30**, 1 (1969).

CORNELIUSSEN, P. E.: Pesticide residues in total diet samples (V). Pest. Monit. J. **4**, 89 ff. (1970).

ELGAR, K. E.: The identification of pesticides at residue concentrations. Adv. Chem. Series **104**, 153 (1971).

FREHSE, H.: Special features in the analysis of pesticide residues: Residue analysis and food control. Residue Reviews **5**, 13 ff. (1964).

——, and G. TIMME: Beatitude through application of latitude. Nat. Meeting, Amer. Chem. Soc., Miami Beach, FL, Sept. 12 (1978); Residue Reviews **73**, 27 (1979).

GUNTHER, F. A.: Current status of pesticide residue methodology. Ann. N.Y. Acad. Sci. **160**, 72 ff. (1969).

—— Pesticide residues in the total environment. Reliable detection and determination,

mitigation, and legislative control and surveillance programmes. Pure & Applied Chem. **21,** 355 ff. (1970).
—— Automation of pesticide residue analysis: Introduction. Proc. Internat. Union Pure & Applied Chem., Tel Aviv, Israel. Vol. IV, 241 ff. (1971).
——, and R. C. BLINN: Analysis of insecticides and acaricides. Numerous pp. New York: Interscience—Wiley (1955).
KAWAR, N. S., G. C. DE BATISTA, and F. A. GUNTHER: Pesticide stability in cold-stored plant parts, soils, and dairy products, and in cold-stored extractives solutions. Residue Reviews **48,** 45 ff. (1973).
SUTHERLAND, G. L.: Residue analytical limit of detectability. Residue Reviews **10,** 85 ff. (1976).
WESTLAKE, W. E., and F. A. GUNTHER: Advances in gas chromatographic detectors illustrated from applications to pesticide residue evaluations. Residue Reviews **18,** 207 (1967).
ZWEIG, G.: The vanishing zero—Ten years later: A status report on recent advances in pesticide analysis. J. Assoc. Official Anal. Chemists **61,** 229 ff. (1978).

Manuscript received April 15, 1979; accepted January 24, 1980.

Effects of pesticides on nontarget organisms

By

George W. Ware[*]

Contents

I. Introduction

Pesticide effects on nontarget organisms have been a source of world-wide contention and concern for more than a decade and are the basis for most legislation aimed at controlling or prohibiting the use of specific

* Department of Entomology, University of Arizona, Tucson, Arizona. Journal Series #2982. Presented in part at the April 1979 US-ROC Cooperative Science Program seminar on "Environmental Problems Associated with Pesticide Usage in the Intensive Agricultural System," Taipei, Taiwan, Republic of China, as sponsored by the National Science Foundation (U.S.A.) and the National Science Council (R.O.C.).

pesticides. For example, the present list of pesticides to undergo Rebuttal Presumption Against Registration (RPAR) numbers 45. All of these are being reviewed because of their effects on nontarget organisms, most with reference to man. Only 13 were triggered by criteria relating to reduction in nontargets and endangered species (birds and feral mammals) while the rest hinged on laboratory animal effects listed as oncogenicity, fetotoxicity, mutagenicity, teratogenicity, and reproductive effects. Thus, most pesticides undergoing RPAR have untoward effects on laboratory animals that are translated as potential effects on man, the ultimate nontarget.

Our most recent significant federal pesticide legislation, the Federal Environmental Pesticide Control Act of 1972, was aimed totally at protecting nontargets—man, his domestic animals and plants, and those not in his domestic dominion. The same is virtually true for all state legislative acts that deal with controlling the kinds and uses of pesticides in commercial and private operations.

Thus it is the nontarget that has actually determined the kinds of laws developed to control pesticides. Because man is the ultimate nontarget, and the most protected, this paper will deal with nontargets other than man.

All taxonomic levels of organisms can be affected by pesticides. Green plants include the angiosperms, gymnosperms, ferns, mosses, and algae; the nongreen are fungi and bacteria. Animals include mammals, birds, reptiles, amphibians, fish, insects, crustacea, worms, coelenterates, and protozoa. All have been shown to be affected by one or more pesticides. There is great variability in species sensitivity to a particular pesticide, as well as great variation in the toxicity of different pesticides to a species. Additionally, for any species, sensitivity to a given pesticide varies with sex, age, nutritional background, stress, health, and the microenvironment. This complexity is important in the evaluation of the precise effect of a pesticide on a species or group of species, but it should not obscure the basic principle that most organisms are affected by some concentration of exposure to one or more pesticides. Our systematic knowledge of the different effects, unfortunately, is very inadequate. Perhaps as a result of complying with the new guidelines for registration of pesticides by the Environmental Protection Agency, our understanding of the broad subject will be increased.

The most readily identified pesticide effects on nontargets were those resulting from the persistent organochlorine insecticides and their metabolites or conversion products on certain species of birds and fish. Significant fish kills have been recorded for DDT, toxaphene, endosulfan, aldrin, dieldrin, and heptachlor used in agricultural and forest pest insect control. Bird kills have been attributed directly to the use of DDT (robins) for elm leaf beetle control and dieldrin for Japanese beetle control, and heptachlor (quail) for Argentine fire ant control in the southern United States. Numerous bird declines have been attributed to food chain accumulations of the chlorinated insecticides and their metabolites.

Pesticides, by necessity, are poisons and would be expected to have adverse effects on any nontarget organism having physiological functions common with those of the target that are attacked or inhibited by the pesticide. These are predictable and dose-related responses. For instance, when a cholinesterase-inhibiting insecticide is applied to a field crop to control one or more pest species, it obviously will also kill other insect species not considered pests. Additionally, avian, mammalian, amphibian, and reptilian species coming in direct contact with the insecticide application may also be affected. Indirect effects can be observed when the insecticide is moved from the application site to another site. There it may be accumulated at several trophic levels to become toxic at the top of a food chain, or reach the secondary site in concentrations that are toxic to nontargets.

A second, and usually unpredictable response, is the effect of the pesticide on dissimilar physiological systems in nontargets. Much of this paper will be devoted to this aspect of nontarget effects.

There are abundant and very thorough reviews on different segments of nontarget effects of pesticides, e.g. wildlife, soil microorganisms, aquatic organisms, etc., and the amount of new knowledge about the effects of pesticides on nontargets is too abundant to make a lengthy review appropriate for this conference. Thus, the author saw his assignment as viewing the subject broadly, pointing to the classic papers where appropriate, and identifying the landmark reviews while providing the reader with a short, general overview of this important aspect of the risk: benefit ratio of pesticide use.

The overall effects of pesticides on nontargets can be categorized as follows: (a) reduction of species numbers, (b) alteration of habitat with species reduction, (c) changes in behavior, (d) growth changes, (e) altered reproduction, (f) changes in food quality and quantity, (g) resistance, (h) disease susceptibility, and (i) biological magnification (PIMENTEL 1971).

II. Biological interactions

There are several broad but identifiable biological interactions of pesticides with organisms that in their early stages of recognition were thought to be unique to pesticides. However, as our knowledge and understanding through research was expanded, it was soon realized that these unusual responses followed precisely established biological principles that were for the most part familiar to the disciple of pharmacology. These interactions include biological magnification or concentration, synergism, potentiation, and liver enzyme induction.

a) Biomagnification

Biomagnification or bioconcentration is the increase or accumulation of some identifiable molecular entity in each step of a food chain. BEVE-NUE (1976) defined bioconcentration as the accumulation of a pesticide in a living organism. Several classic examples occur in the persistent, lipid-soluble organochlorine insecticides. There are none found in the organophosphate or carbamate classes of insecticides. Probably the earliest documentation of biomagnification of a persistent organochlorine insecticide is the Clear Lake, California, episode. In 1949, to control a nonbiting gnat, DDD (TDE) was applied to the lake, its breeding site. DDD was chosen instead of DDT because of its lower fish toxicity. The spraying was quite successful until 1951, when the gnats reappeared. Then each summer routine applications were made to the lake, sometimes more than once. In 1954 large numbers of western grebes, a diving bird, were found dead in the lake area, and their reproduction was reduced noticeably (HUNT and BISHOFF 1960).

To understand this phenomenon, we must examine the various levels of DDD in water and various organisms. The Clear Lake water contained 0.02 ppm DDD; plankton, 5 ppm; small fish, 9 to 10 ppm; predatory fish $\simeq$2,000 ppm; and grebes >2,000 ppm. The dead grebes had accumulated lethal levels of DDD in the brain (HUNT and BISHOFF 1960).

The declining reproduction observed in the Clear Lake episode was attributed to the indirect effect of concentrated DDD on eggshell breakage during incubation. This was probably the direct effect of DDD on calcium metabolism similar to the DDT effects on avian reproduction (WARE 1975).

A second and similar example, due not to direct application, but rather due to agricultural runoff and lake water contamination took place in 1965 in Lake Michigan (HICKEY et al. 1966). In that instance water measurements of organochlorine insecticides were not detectable. However, bottom sediment contained 0.01 ppm; small invertebrates (plankton) 0.41 ppm; fishes, 3 to 8 ppm; while Herring gulls, the apex of the food chain, contained 3,177 ppm of organochlorine insecticides.

Biomagnification has been identified with several of the organochlorine insecticides, and more specifically with those that are not readily metabolized by vertebrates, e.g. DDT, DDD, dieldrin (aldrin), heptachlor (and heptachlor epoxide), and mirex. Ultimately, biomagnification has led directly to the death of the top of food chains, or indirectly to reproductive failures.

A well-documented instance of biomagnification of DDT in a marsh ecosystem was published by WOODWELL in 1967, the data for which are presented in Table I.

Once the persistent material reaches the water, it is quickly absorbed by microorganisms. The degradation of DDT and its biological magnification by freshwater invertebrates, was studied by JOHNSON et al. (1971), using ^{14}C-labeled DDT. The uptake and magnification by all

Table I. *DDT residues in biota samples from a
Long Island marsh area* (WOODWELL 1967).

Substrate	PPM (wet wt, whole-body basis)
Plankton	0.04
Water plants	0.08
Marsh plants	
shoots	0.33
roots	2.80
Cricket	0.23
Diptera	0.30
Fluke	1.28
Mud snail	0.26
Clam	0.42
Bay shrimp	0.16
Eel	0.28
Silversides	0.23
Minnow	0.94
Fundulus	1.24
Blowfish	0.17
Billfish	2.07
Green heron	3.57
Tern	3.15–6.40
Osprey (egg)	13.8
Merganser	22.8
Cormorant	26.4
Gull	3.52–75.5

of the invertebrates tested were direct and rapid. Within 3 days from
the beginning of exposure, the magnification factor was impressive:

Organism	Magnification factor
Cladocera	25,400 to 114,100
Amphipoda	4,600 to 20,600
Decapoda	880 to 2,900
Ephemeroptera	9,400 to 32,600
Odonata	910 to 3,500
Diptera	7,800 to 133,600

It is apparent from these data and their relationship to higher links
of the food chain shown in Table I, that aquatic invertebrates are the
key factor in biomagnification.

b) Synergism

Synergism is usually considered to be the enhancement of toxicity
of a toxicant to an organism by another compound that would normally
be considered nontoxic or relatively nontoxic. Combinations of unlike

classes of pesticides, e.g. insecticides and herbicides or fungicides and insecticides, potentially offer enhanced or inhibited activity. The literature contains a host of pesticide-pesticide and pesticide-pollutant interactions resulting in increased biological effects. LIANG and LICHTENSTEIN (1974) demonstrated that ethyl parathion and p,p'-DDT were synergized with the herbicide atrazine against mosquito larvae (*Aedes aegypti*) in turbulent water and against fruit flies (*Drosophila melanogaster*) exposed to pesticide-treated soil.

Other forms of synergism with different classes of pesticides have been observed in fish. Bluegills tested with 37 combinations of copper sulfate, DDT, carbaryl, parathion, malathion, and methyl parathion resulted in additive effects for 22 combinations, synergism for 13, and antagonism for 2. Most of the combinations showing synergism contained one or two organophosphates (MACEK 1969). WEISS (1959) found EPN and malathion to be synergistic in fish, while others found DDT and 2,4-D to be synergistic (CAIN 1965).

Many commercial herbicide, insecticide, and fungicide formulations contain more than one basic chemical. A parallel problem with synergism may come from the routine application in agricultural areas of more than one pesticide in quick succession in a single season. Thus, knowledge of synergistic effects on birds, fish, and mammals in specific ecosystems is essential for effective resource management and wildlife protection.

c) Potentiation

Potentiation is the greater-than-additive toxicity of two organophosphate insecticides, the extent known for this phenomenon. The first case of marked potentiation of toxicity among organophosphates was with EPN and malathion in the dog and rat (FRAWLEY *et al.* 1957). In this phenomenal case $1/40$ LD_{50} of malathion plus $1/50$ LD_{50} of EPN gave 100% mortality!

The mechanism of potentiation has been demonstrated to be the inhibition by one toxicant of the enzyme(s) responsible for the degradation of the other (SEUME and O'BRIEN 1960). Antagonism, likewise, is explained as the inhibition by one toxicant of the enzyme(s) involved in the activation of the other.

Insects, because of their inherent carboxyesterase activity, have demonstrated only slight response to combinations of organophosphates, except where resistance due to increased levels of this enzyme are involved (PLAPP *et al.* 1963), specifically in mosquitoes and houseflies.

Potentiation was observed by FABACHER *et al.* (1976) between the cotton defoliant DEF® (S,S,S-tributyl phosphorotrithioate) and methyl parathion to the mosquitofish (*Gambusia affinis*). In this case there was no mortality in fish treated with 0.5 ppm DEF, while fish exposed to 5.0 ppm methyl parathion resulted in only 8% mortality. However, when combined at the same concentrations there was 89% mortality, or a

10-fold potentiation. Because DEF is more toxic than methyl parathion to the fish, it appears that DEF is potentiated by methyl parathion.

How this relates to nontarget organisms is purely speculative at this time. It is conceivable that combinations of organophosphates applied to crops could exhibit potentiation (or antagonism) in field rodents, birds, and to a lesser extent in insects, resulting in mortality of feral vertebrates and notable success (or failure) in insect pest control.

d) Liver enzyme induction

Organochlorine insecticides are known to induce the production of hepatic detoxifying enzymes in mammals, birds, and fish at extremely low levels of intake. This phenomenon has the potential for influencing several effects of these insecticides in the general environment. One important aspect that may be so influenced is the antagonism of insecticide residue storage in animal tissues due to accelerated metabolic degration. A result may be lowered chronic toxicity of certain insecticide combinations. The presence of many organochlorine pesticides in the environment offers numerous opportunities for such interaction in various species. Too, disturbed endocrine relationships due to increased hormone turnover by induced enzymes may lead to physiological aberrations, such as that suggested by PEAKALL (1967) as the basis for egg fragility and poor reproduction in certain birds. The thin eggshell aspect is discussed elsewhere in this paper.

Cyclodiene epoxides have the lowest apparent threshold (1 ppm) for inducing hepatic enzymes, the value for DDT is slightly higher, while other types (heptachlor, toxaphene, aldrin, methoxychlor) are not active below about the 5 ppm level (STREET et al. 1970). The antagonism of dieldrin storage in rat adipose tissue by DDT, an effect attributed to induced enzyme activity, also seems to occur at about the same dosage of DDT as do the measured enzyme responses.

Certain birds, fish, and mammals, however, may be exposed to dietary levels of organochlorine insecticides exceeding these thresholds. Reports indicate organochlorine insecticide residues in diets of birds range up to 10 ppm (DUGGAN and WEATHERWAX 1967). As a result, the probability still exists that significant enzyme induction by residual insecticides is still underway in some wildlife species as a result of persistent organochlorine residues in the environment.

III. Wildlife

Much has been written with respect to the effects of pesticides on nontarget wildlife. It was this topic that inspired the late RACHEL CARSON to write her highly influential *Silent spring* in 1962. Her charge, that man was poisoning the earth and himself by his efforts to control pests with chemicals, created international controversy.

Some of the better reviews of pesticide influences on wildlife are

180   G. W. WARE

by COPE (1971), NEWSOM (1967), BROWN (1978), MATSUMURA et al. (1972), STICKEL (1973), and by far the most comprehensive by PIMENTEL (1971). There are many others, but the basic subject matter is well covered in those mentioned.

Traditionally, the greatest and most significant effects on wildlife were wrought by large-scale applications of organochlorine insecticides on fish and birds. The chlorinated insecticides are astonishingly toxic to fish. Acute effects were so obvious and so drastic that it was apparent from the beginning of their use that these insecticides could not be used indiscriminately in agriculture or in public health activities. Trout and salmon were the two most reported on, and age classes were affected differentially with severity inversely related to age (KEENLEYSIDE 1959).

In birds, two effects were witnessed, but not recognized as distinct at the outset. The first was mortality due to the accumulation of organochlorine residues in their food, eventually resulting in lethal brain levels of the parent insecticide (STICKEL et al. 1966 and 1969).

The second effect was death of the embryos due primarily to premature cracking of thin egg shells and secondarily to lethal pesticide levels in the embryos (HICKEY and ANDERSON 1968). The thin eggshell syndrome probably applies to documented population decreases mainly in predatory birds that feed on fish or small birds. These include the bald eagle (Haliaeetus leucophalus), the osprey (Pandion haliaetus), the peregrine falcon (Falco peregrinus), the European sparrow hawk (Accipiter nisus), and the brown pelican (Pelicanus occidentalis). These population declines have been attributed to thin eggshells, which break during turning and normal incubation activities of the parents. Current theories that attempt to explain this phenomenon are (a) reduced levels of carbonic anhydrase, (b) inhibited calcium metabolism, (c) induction of hepatic microsomal enzymes that metabolize steroid hormones (WARE 1975), (d) hypothyroidism, and (e) ATP-ase inhibition (BROWN 1978).

Deaths of mammals, reptiles, and amphibia have not drawn attention equal to that of birds and fish, and consequently have not been intensively studied. It is known, however, that frogs, toads, and their tadpoles are considerably less susceptible to insecticide poisoning than fish, while snakes as a group appear to be about as susceptible to pesticide intoxication as the amphibia (BROWN 1978).

Bats are probably the most susceptible mammal to DDT poisoning. The oral LD_{50} for the big brown bat, Eptesicus fuscus, is 30 mg/kg compared to 400 mg/kg for laboratory mice (LUCKENS and DAVIS 1964). As a consequence both DDT dusts and wettable powder sprays are used to eliminate bat colonies suspected of carrying rabies (BROWN 1978).

The indirect effect on wildlife by pesticides is a subject not thoroughly investigated but certainly of consequence. Herbicides often indirectly affect fish populations by destroying aquatic vegetation. In lakes or ponds with little or no circulation, significant amounts of CO_2 may be released when aquatic plants fall to the bottom and decompose, killing fish and other aquatic life. In other situations, pesticides applied to ponds may deplete the oxygen in the water within a few hr, jeopardizing fish survival. Just as chronic toxicity may be more damaging to a population than is acute toxicity, indirect effects on wildlife may often have more serious consequences than the direct influences of sublethal exposures.

IV. Domestic animals

Livestock do not normally come in contact with pesticides in amounts sufficient to cause intoxication, except in an occasional accident. As in other animals, they are exposed to sublethal doses and accumulate detectable quantities of persistent insecticides in their tissues.

During the days of heavy calcium arsenate use on cotton, 1925 to 1948, aerial application of these dusts occasionally resulted in drift to adjacent pastures. Dairy animals feeding in these pastures sometimes ingested lethal levels. Similarly, a few such incidents relating to cows, horses, and other grazing animals have been attributed to the organo-chlorine insecticides.

The role played by livestock in the transfer of pesticide residues to human beings is one of concern because it is in such products as meat, milk, and eggs that most of the insecticides in the human diet are found (CAMPBELL et al. 1965). For instance, in Arizona DDT was banned from all agricultural uses in 1969 because of forage contamination from DDT application to cotton, resulting in milk residues of DDT and DDE in excess of tolerance. Substantial residues of DDE were also measured in beef being fattened in feedlots on feeds grown in Arizona (WARE 1974).

In 1979 the insecticide toxaphene was banned from all but a few minor uses in Arizona for the same reasons (WARE 1979). Toxaphene applied by air to cotton in combination with methyl parathion and other insecticides, drifted onto alfalfa, which was fed either as hay or green-chopped fodder to dairy animals. Over the last 3 yr some 30% of all Arizona dairies were reported to have toxaphene residues in whole milk in excess of tolerance (0.05 ppm) one or more times.

The accumulation of dieldrin and heptachlor epoxide at levels of 11 ppm in adipose tissue and 13 ppm in butterfat (ca. 0.5 ppm in whole milk) has been reported in cattle exposed to rangeland treated at 2 oz/A (ANONYMOUS 1959). BRUCE et al. (1965) have shown that the lower the concentration in the diet of dairy animals the higher the percentage stored in adipose tissue and excreted in milk.

Chickens are not unlike cows in this respect. Eggs have been found to contain residues of all chlorinated insecticides when fed in their diets, and inadvertently produce contaminated eggs by absorption of lindane or BHC through their feet from roosts or litter treated for louse control (WARE and NABER 1961).

V. Agricultural arthropods

a) Predators and parasites

Parallelling the effects of pesticides on wildlife are the effects of insecticides on beneficial arthropods interspersed with the crop insect pests. Most of the insecticides in use today have broad-spectrum effects. They are lethal to a wide range of insects and other invertebrates, including beneficial competitors, predators, and parasites of the target pest insects. When insect pest populations are drastically reduced, their natural enemies are reduced even more. A resurgence in the pest population can then occur, resulting in increased damage to the crop.

Insecticides are applied to a crop for only a few pest species. In most cases, these key species require chemical control to prevent economic losses. The history of insecticide usage, however, illustrates that additional problems are created, either by the rapid resurgence of the treated pest or by raising minor pests to the role of secondary or major pest status.

It may be argued that insecticides affect both pests and natural enemies alike and the capacity of the pest species to resurge should apply to the enemy populations as well. This is true, and some beneficial insects and mites have developed resistance to insecticides, along with the pest species (DEBACH and BARTLETT 1951). However, natural enemy populations suffer double jeopardy under these conditions. First, they are subjected to lethal levels of the insecticide along with the pests, and are reduced to relatively low numbers. Second, the surviving natural enemies are left with little to feed upon and may further decline because of starvation or may abandon the area in search of food.

Two good examples will illustrate that insecticides reduce natural enemies with a resulting increase in pest populations. In 1946 to 1947, thousands of acres of California citrus developed damaging populations of the cottony-cushion scale following applications of DDT for other pests. This was caused by elimination of the vedalia beetle, a predator responsible for controlling the scale since its introduction in 1888 from Australia. It has since been demonstrated experimentally that elimination of the vedalia was the sole cause of the resurgence of cottony-cushion scale (DEBACH 1947).

A second example illustrates conclusively that elimination of a predator may permit resurgence of a pest species. Entomologists have shown that removing the predatory mite, *Typhlodromus reticulatus*, with parathion treatments for cyclamen mite control on strawberries, permits the cyclamen mite to increase 15 to 35 times whereas cyclamen mite populations continue to decline gradually where the predaceous mite is left undisturbed.

Systemic insecticides can have deleterious effects on predator populations thus influencing a resurgence of the pest species. RIDGWAY et al. (1967) sampled populations of arthropod predators, certain parasitic wasps, and the cotton bollworm/tobacco budworm complex, following in-furrow applications of the 4 most commonly used systemic insecticides on cotton. Populations of predators, particularly the Hemipterans, were reduced as were spiders and wasp populations, but to a lesser extent, while eggs and larvae of the pests increased.

Another example of resurgence was seen in pecan orchards in southern Arizona. Treatment of pecan trees periodically during spring and summer with the systemic insecticide aldicarb (Temik) suppressed the black-margined yellow aphid (*Monellia caryella*) throughout the growing season, and because of a lack of prey the green lacewing (*Chrysopa carnea*) population was nonexistent. The untreated trees, however, had a continuous, but noneconomic infestation of aphids throughout the season with a balanced population of green lacewings. In late August the aldicarb had lost its effectiveness and the aphids increased to high levels of economic significance, in contrast to the untreated trees that continued to support low to moderate aphid infestations. In this instance, the predators were indirectly affected by the systemic treatments due to lack of available food (HUBER 1979).

Insecticides may be more selective in killing predators than the pests, as demonstrated in recent Arizona studies. The tobacco budworm *Heliothis virescens*, has just recently become a serious pest of cotton in Arizona, and at the same time rapidly became resistant to the commonly used methyl parathion. Populations collected in 1972 from cotton had an LD_{50} of 27 μg. In 1978 field populations reached a high level of resistance showing an LD_{50} of 212 μg, or a 10$\times$ resistance (WATSON 1979). At the same time its primary predators were also tested to determine their tolerance to methyl parathion expressed as LD_{50}s: *Nabis alternatus*, 2.1 μg; *N. americoferus*, 1.4 μg; *Geocoris pallens*, 2.0 μg; and *Orius tristicolor*, 0.65 μg (Table II) (PAPE 1979).

Table II. LD_{50}s of methyl parathion (μg/g).

Nabis alternatus	2.1
Nabis americoferus	1.4
Geocoris pallens	2.0
Geocoris punctipes	2.0
Orius tristicolor	0.7
Heliothis virescens	212

This clearly demonstrates that a single application of methyl parathion to a cotton field would reduce the budworm population only slightly, while virtually eliminating its most important predators. In this instance, however, a greater degree of susceptibility (100-fold) in the predators to the insecticide than their host could explain a pest control failure or resurgence.

In a recent study concerned with one of the "third generation" insecticides, 7 insect predators were monitored in cotton fields treated with the chitin inhibitor diflubenzuron, fields treated with conventional insecticides, and an untreated field. Nine applications of diflubenzuron at 2 oz/A, compared to the untreated, established a trend toward a slight adverse effect when all species or groups were considered. The bigeyed bug, *Geocoris punctipes*, was the only species showing significantly reduced reproduction. Lady beetles, *Coleomegilla maculata*, also had significantly lower adult populations. This was in marked contrast to the highly adverse effects of 7 applications of toxaphene-methyl parathion, 2.1 lb a.i./A, upon the predator populations (KEEVER *et al.* 1977).

b) Minor pests become major pests

The history of insecticide usage has illustrated 2 major population influences in treated crops, (1) the rapid resurgence of the exposed pest population, and (2) the raising of minor pests to the role of secondary or major pests.

Some pests, such as the cotton leafworm and cotton aphid, have declined in pest status following the use of organophosphates for control of organochlorine-resistant boll weevils. Other cases can be cited. However, the reverse is more generally the case. That is, formerly minor pests have become major pests. Without question, the suppression of natural enemies by insecticides is the main reason for these switches (WATSON *et al.* 1976). For instance, in Arizona, spider mites, cotton leafperforator, and the tobacco budworm were classed as minor pests of cotton until about 1972 to 1974, when the increased destruction of their predators by intensive use of organophosphates permitted these to become major pests, particularly the budworm and leafperforator (WATSON 1979).

Another classic case of elevating minor pests to major pest status occurred in Central America. In 1950, when organochlorine insecticides were introduced there, cotton had only 2 pests of economic importance, the boll weevil (*Anthonomus grandis*) and the cotton leafworm (*Alabama argillaceae*). At that time only DDT, BHC, and toxaphene were being used (ICAITI 1976). Yields increased, as did the number of pests. Three species, not then recognized as pests, emerged from the continued insecticide applications: the cotton bollworm (*Heliothis zea*), the cotton aphid (*Aphis gossypii*), and the false pink bollworm (*Sacododes pyrolis*).

Attempts to control these 5 pests elevated 5 new ones from minor to major pest status. These were 2 armyworms (*Prodenia* spp, and *Spodoptera* spp.), whiteflies (*Bemisia tabaci*), cabbage loopers (*Trichoplusia ni*), and plant bugs (*Creontiades signatum*). During this interval, the applications needed per growing season increased from 5 in 1950 to 28 in the mid 1960's (Table III).

Table III. *Minor pests become major pests—Central American cotton.*

1950	1955	1960's
	Pesticide applications/season	
0–a few	8–10	28
boll weevil	boll weevil	boll weevil
leafworm	leafworm	leafworm
	bollworm	bollworm
	cotton aphid	armyworm (2 species)
	false pink bollworm	whitefly
		cabbage looper
		plant bug

BOTTRELL and RUMMEL (1978) reported on recent attempts to reduce prehibernating populations of the cotton boll weevil in west Texas in the fall with application of malathion and azinphosmethyl. These produced an immediate ecological disruption as evidenced by the increase of the nontarget pests, cotton bollworm and tobacco budworm (*Heliothis zea* and *H. virescens*). At peak densities, *Heliothis* populations were 4- to 12-fold greater in treated than in untreated fields and commonly reached higher levels in cotton fields treated with azinphosmethyl. No carry-over effect of the late-season buildup was detected in the following spring and summer.

Spider mite build-up has frequently occurred in cotton treated repeatedly with organochlorine or certain organophosphate insecticides. LEIGH (1979) has reported that methyl parathion or phosphoric acid applications to cotton stimulated mite egg production 25 to 35% during the first 8 to 14 days of mite development. The total egg production was the same as in the controls, only earlier, giving rise to shorter intervals between generations.

c) Resistance in nontargets

The development of resistance in pests of agricultural and public health importance indicates that resistance must occur also in beneficial insects, and it must be widespread. There are several instances in which nontarget insects, some of which may be pest species, have developed resistance to insecticides intended for other pests. The long and intensive use of agricultural insecticides in Central America caused insecticide resistance in the principal malaria mosquito vector, *Anopheles albimanus*. It was determined later that the mosquitoes had become resistant to DDT, dieldrin, malathion, and propoxur as a result of their agricultural use rather than from vector control (Table IV) (BROWN and PAL 1971).

Very recent studies in Arizona (SHOUR and CROWDER 1979) indicated that the green lacewing (*Chrysopa carnea*), an important predator in cotton fields, was apparently not affected by applications of insecticides in the field, including the synthetic pyrethroids. When brought

Table IV. *Resistance in Anopheles albimanus* (1974).

	DDT (Guatemala)	Propoxur (El Salvador)
Cotton areas	79%	35.4%
Noncotton	53.6%	6.9%

to the laboratory both Arizona and California strains of the lacewing as larvae were quite tolerant to the pyrethroids permethrin (Pounce®) and fenvalerate (Pydrin®). However, less than 2% of the larvae pupated from the fenvalerate treatment, while those exposed to permethrin survived.

A wasp parasite of the boll weevil, *Bracon mellitor*, was demonstrated to have 4-fold resistance to DDT, carbaryl, and methyl parathion and 8-fold resistance to a DDT-toxaphene combination after only 5 generations of exposure in the laboratory (ADAMS and CROSS 1967). Additional research will show insecticide resistance to be common in populations of predators, parasites, pollinators, scavengers, and species of importance in food chains.

VI. Pollinators

United States agriculture requires honey bees for pollination of some 55 of the more than 200 crops grown to produce commercial quantities of seeds and fruits, while about 1 billion lb of pesticides are used annually to control crop pests (BERRY 1978).

Inadvertently, pesticide (specifically insecticide) applications destroy hundreds of thousands of domestic honey bee colonies in the United States each year (see Tables V and VI). This inflicts serious economic

Table V. *Bee colonies and insecticide use in Arizona.*

Year	Colonies (1000's)	Chlorinated hydrocarbons (1,000 lb)	Organo- phosphates (1,000 lb)	Carbamates (1,000 lb)
1965	114	1,764	461	20
1971	53	2,342	5,124	372

losses both to beekeepers and to growers whose crops depend on bee pollination (alfalfa seed, almonds, apples, apricots, and all crops grown for seed). Impossible to measure are the losses of wild, solitary bees which are generally more efficient in pollination than honey bees. The same factors that are detrimental to honey bees are equally if not more detrimental to wild pollinators. Included are the use of herbicides and other forms of weed control which further reduce bee forage.

Most pesticides are not hazardous to bees (ATKINS *et al.* 1973). Of

Table VI. *California bee colony losses.*

Year	Colonies of bees	Pesticides	Other losses	Total
1969	537,000	82,000	117,000	199,000
1970	521,000	89,000	70,000	159,000
1971	511,000	76,000	32,000	108,000
1972	500,000	40,000	30,000	70,000
1973	500,000	36,000	31,000	67,000
1974	500,000	54,000	33,000	87,000

399 pesticides 20% are highly toxic, 15% moderately toxic, and 65% relatively nontoxic or nontoxic to honey bees. In general, the insecticides are the most toxic group headed by the carbamates and organophosphates. The acaricides, fungicides, and herbicides are relatively nontoxic (ATKINS *et al.* 1977).

Two good examples of honey bee losses due to insecticide kills occurred in Arizona and California. In 1965 Arizona was almost totally dependent on the chlorinated insecticides (Table V) and supported 114,000 colonies of bees. In 1971, with the change to organophosphate insecticides after the 1969 ban on DDT, the colony numbers dropped to 53,000 or a 50% loss (WARE 1973). In that same period California was losing more than 80,000 colonies annually to pesticides (Table VI), again due in great part to the shift from organochlorine to organophosphate insecticides (FLINT and VAN DEN BOSCH 1977).

Carbaryl has been frequently mentioned as a particularly harmful insecticide to bees. In the laboratory it is less toxic to adult bees than several other commonly used insecticides, but pollen contaminated with carbaryl may remain toxic to bees for more than 10 wk. Young bees feed heavily on pollen and if it is contaminated with carbaryl, young bees that eat it die and accumulate near the hive entrance (MOFFETT *et al.* 1970). This makes the kill more obvious and prolonged than that caused by faster acting insecticides which kill bees in the field. Ultralow-volume (ULV) applications of the undiluted technical malathion used in spraying large acreages have also caused serious bee kills (LEVIN *et al.* 1968).

Formulations make a great difference in an insecticide's toxicity to honey bees. Dusts are more hazardous than sprays. Wettable powders are usually more hazardous than emulsifiable concentrations or water-soluble formulations. Micro-encapsulated materials are more toxic than all other formulations because the vinyl spheres are about the size of pollen grains and are treated as such, leading to brood kills (ATKINS *et al.* 1977).

Fine sprays are less toxic than coarse sprays, and ULV formulations (undiluted) are more toxic than diluted sprays. Granulars are the safest of all formulations. Furthermore, night applications are less destructive than daylight applications, while ground applications cause fewer losses than aerial applications (ATKINS *et al.* 1977).

Most beekeepers occasionally experience bee poisoning from pesticides, and the source of the poisoning is usually difficult to determine. Some sections of the United States have had particularly serious bee losses. As an example, the spraying of sweet corn with sevin (carbaryl) for the control of corn earworms was for several yr the main cause of an annual loss approaching 25% of commercial colonies in Washington state. The problem is being eased by turning to other insecticides. After many beekeepers were forced out of business, particularly in Arizona and in parts of California (Tables V and VI), state and national organizations sought government help. Following a period of aggressive lobbying, the federal government developed a "Beekeeper Indemnity Payment Program" which took effect in July, 1971. Under the program, beekeepers who suffered losses caused by pesticides used as recommended and labeled could be monetarily reimbursed (*U.S. Department of Agriculture* 1971).

Beekeepers now look upon pesticide poisoning as one of their most serious problems. Even though much research is being carried on to find nonchemical methods of pest control, increased use of pesticides is generally predicted. The answer lies eventually in the development of thoughtful, integrated, pest-management systems which rely on a combination of all ecologically sound control methods.

VII. Soil organisms

a) Arthropods and annelids

Insecticides, nematicides, molluscicides, fungicides, and herbicides in soil are potentially able to change populations of soil invertebrates, either directly or indirectly. Because insecticides, nematicides, and molluscicides are selected for their toxicity to invertebrates they have the greatest direct effect. Herbicides may strongly influence soil invertebrate populations indirectly by their effects on vegetation which provides food and habitat for many of the animals, and it seems probable that when fungicides kill microorganisms they alter the food supply of the many invertebrates that feed on these. Some insecticides, particularly the organochlorine compounds, accumulate in the fatty tissues of animals; thus, if these chemicals occur in invertebrate tissues, they are potentially hazardous to animals further up food chains, particularly predatory vertebrates (EDWARDS and THOMPSON 1973).

Soil arthropods obviously receive a wide range of exposures to a wide variety of pesticides, some within an individual's life span. Gross effects are difficult if not impossible to establish without a tremendous background of research with individual compounds in different soil types. Because several wide-ranging reviews have been published, particularly by EDWARDS and THOMPSON (1973), an outstanding contribution, it would be redundant to reassess this subject.

Most of the organochlorine insecticides are characteristically stable in soils, and should exert their toxic effect for an extended period, though selectively, over the wide variety of soil arthropods. Work done in Arizona (DRAKE *et al.* 1971) indicted that 80 to 90% of the soil mites were oribatids in plots treated with DDT, endosulfan, and endrin, while 70 to 80% of those in the Strobane-DDT treated plot were mesostigmatids. In the control neither group would comprise more than 30%. Ground pearl nymphs were obtained from all except the Strobane-DDD plot. The malathion-perthane plot yielded 24% more animals than any other, including the controls. These data show only that insecticides do indeed shift species populations. The significance of such shifts remains to be identified.

Interpreting this kind of information is extremely difficult, since it is not known if removal of competition and predation are the only causes of increased numbers. EDWARDS *et al.* (1967) noted that neither aldrin nor DDT greatly changed numbers of soil arthropods.

Earthworms were reduced by 43% from 37.2 lb/A of DDT (POLIVKA 1951), while 5 lb/A of aldrin (POLIVKA 1953) significantly reduced earthworms when applied to golf courses. Chlordane has commonly been recommended for earthworm control in turf and lawns throughout the United States (RAW 1963). Recent work (WRIGHT 1977) indicates that earthworms are particularly susceptible to benomyl and the newer systemic fungicides.

b) Microorganisms

Soil microorganisms are frequently the major and sometimes the only means by which pesticides are eliminated from a variety of ecosystems, and are therefore important in governing persistence. A number of products are formed during degradation, some of which may be persistent and several of which may be quite toxic. If an intermediate is both persistent and toxic, a major environmental hazard may appear. That a pesticide is persistent suggests that it is unsuitable as a substrate for microorganisms in that particular site.

Our concern with the persistence of organochlorine insecticides and certain herbicides arises because of the invalidity of the early view that soil and water are perfect waste disposal media. It is evident that we cannot continue to dump a diversity of organic compounds into these microbial habitats and assume that these substances will be degraded.

As with the section on wildlife, there are many good and complete reviews on the effects of pesticides on soil microorganisms (BOLLAG 1972, PFISTER 1972, EDWARDS 1973, BROWN 1978, ALEXANDER 1972, PIMENTEL 1971, LAVEGLIA and DAHM 1977, KUHR and DOROUGH 1976, BROOKS 1974, SETHUNATHAN 1973, JOHNSEN 1976, TU and MILES 1976) to which the reader is referred. The most complete review, however, is by HELLING *et al.* (1971).

A comprehensive study on the effect of organochlorine, carbamate, and organophosphate insecticides on microbial respiration in soils was reported by BARTHA *et al.* (1967). It was proposed that mechanisms responsible for changes in microbial respiration in soils treated with insecticides may be combinations of: (a) uncoupling of oxidative phosphorylation, (b) the nontoxic parent compound being converted to a stable, toxic product, (c) or an insecticide with selective toxicity inhibits CO_2 production of some microbes while being oxidized by others that are resistant to its selectivity. It was emphasized that concentrations of pesticides greatly in excess of those commonly used in agriculture were required to produce effects, and that additional organic matter reduced toxicity of pesticides and increased microbial degradation of the toxicants.

Because the subject is one of great magnitude with many fine reviews, the author will not attempt to rework the literature.

A generalized overview of the soil half-lives of the various classes of pesticides would occur somewhere near the following order (adapted from BROWN 1978):

Insecticides:	Organochlorines	10s of years
	Organophosphates	months
	Carbamates	weeks
Herbicides:	Urea, triazine, picloram, benzoic acid, and amide	10s of months
	Phenoxy, toluidine, and Nitrilex	months
	Carbamate and aliphatic acid	weeks
Fungicides:	Volatile	days
	Less volatile	weeks to months

The literature cited above provides extensive coverage on the persistence of a wide range of insecticides and herbicides. Unfortunately, less is known about the longevity in soil of fungicides. A list of genera of microorganisms known to metabolize pesticides is presented in Table VII (WARE and ROAN 1971).

VIII. Aquatic microorganisms and plankton

Many of the same soil particles, microorganisms, and pesticides found in soils are also found in freshwater and estuarine ecosystems, and similar relationships may exist. Aquatic microorganisms are in many instances "weeds" in that they are merely soil microorganisms out of place. Thus, much of the biological activity taking place in aquatic environments can also be found in soils (WARE and ROAN 1971).

Aquatic microorganisms and plankton are those microflora and microfauna commonly found in freshwater, brackish, and marine environ-

Table VII. *Genera of selected microorganisms known to metabolize pesticides.*

Bacteria	Actinomycetes	Fungi	Algae	Yeasts
Achromobacter	Mycoplana	Acrostalagmus	Chlamydomonas	Saccharomyces
Aerobacter	Nocardia	Aspergillus	Chlorella	Torulopsis
Agrobacterium	Streptomyces	Clonostachys	Cladophora	
Alcaligenes		Cylindrocarpon	Cylindrotheca	
Arthrobacter		Fusarium	Oscillatoria	
Azotobacter		Geotrichum	Vaucheria	
Bacillus		Glomerella		
Clostridium		Helmintho-		
Corynebacterium		sporium		
Escherichia		Mucor		
Flavobacterium		Myrothecium		
Klebsiella		Penicillium		
Micrococcus		Stachybotrys		
Paracolobactrum		Trichoderma		
Proteus		Xylaria		
Pseudomonas				
Rhizobium				
Sarcina				
Serratia				
Sporocytophaga				
Thiobacillus				
Xanthomonas				

ments. Plankton are organisms found floating or drifting almost passively and are carried about by wave action and currents. Overall the effects of pesticides on these organisms are as varied and broad as the effects of pesticides on terrestrial organisms.

Pesticides enter aquatic environments mostly unintentionally. WEST-LAKE and GUNTHER (1966) classified environmental contamination either as direct or intentional application and as indirect or unintentional. Table VIII is a modification of their classification system.

Table VIII. *Sources of pesticides in the aquatic environment.*

A. Intentional introductions
   1. Pest control of objectional flora and/or fauna
   2. Industrial wastes
     a. Pesticide manufacturers and formulators
     b. Food industry
     c. Moth-proofing industry
   3. Disposal of unused materials
   4. On-site field cleaning of application, mixing, and dipping equipment
   5. Disposal of commodities with excessive residues
   6. Decontamination procedures
B. Unintentional introductions
   1. Drift from pesticide applications to control objectional flora or fauna
   2. Secondary relocation from target area via natural wind and water erosion
   3. Irrigation soil water from target areas
   4. Accidents involving water-borne cargo
   5. Application accidents involving missed targets or improper chemicals

A detailed recitation of the pesticides detected in the aquatic environment at various places and times would include all pesticides that have been or are in current use. The persistence of various pesticides in the nonliving components of the aquatic environment must be recognized particularly where this persistence may be greater than in the soil. A specific example, from SCHWARTZ (1967), of the metabolism of 2,4-D being much slower in aqueous than in soil environments is illustrative of this situation.

Toxicity is usually regarded only as the lethal effects of the chemical upon the organisms. Other direct effects that should be considered are changes in growth and specific metabolic rates, *i.e.*, photosynthesis.

The water solubilities of most pesticides, particularly the organic insecticides, are generally quite low but vary from about 1 ppb to complete miscibility with water. Their relative solubilities from least to greatest are the organochlorine, carbamate, and organophosphate insecticides (GUNTHER *et al.* 1968).

Partly because of their low water solubilities, pesticides have an affinity for living organisms. Microorganisms and plankton quickly accumulate quantities of pesticides from the water medium and retain them in and on their tissues. Aquatic fauna of all sizes tend to remove pesticides from the water and store them (COPE 1966), thus beginning the first step of biomagnification or bioconcentration.

UKELES (1962) presented data on the toxicity of 17 pesticides to 5 species of marine plankton, 12 of which are displayed in Table IX. His data have been rearranged to indicate the highest concentrations preventing any growth of the cultures during a period of 10 days, and indicate that of the insecticides tested only toxaphene approaches the phytotoxicity of herbicides tested against marine phytoplankton.

Table IX. *Pesticide concentrations* (ppm) *preventing growth of marine phytoplankton.*

Pesticide	Proloccus sp.	Chlorella sp.	Dunaliella euchlora	Phaco-dactylum tricor-nutum	Mono-chrysis lutheri
Dipterex (trichlorfan)	1,000	500	500	500	100
TEPP	500	500	500	500	500
Phenol	500	500	500	100	100
Orthodichlorobenzene	13	13	13	13	13
Sevin (carbaryl)	10	10	10	<0.1	1
Nabam	10	10	1	1	1
Lindane	>9	>9	>9	7.5	7.5
Toxaphene	0.15	0.07	0.15	0.04	0.01
DDT	>60	>60	>60	>60	60
Fenuron	29	2.9	2.9	2.9	2.9
Monuron	0.02	0.02	0.02	0.02	0.02
Diuron	0.004	0.04	0.004	0.004	$<2 \times 10^{-5}$

Selected data from Cope's (1966) investigations are presented in Table X. As might be expected, the insecticides are much more toxic to *Daphnia* than are the herbicides.

Because of the appearance of the polychlorobiphenyls (PCBs) in the analytical literature as contaminants which may be mistakenly identified as some insecticide residues, it is not reasonable to assume that all residues of DDT and related metabolites reported in the literature on aquatic contamination are in fact as stated. Schechter (1969) is of the opinion that the PCBs are primarily a problem in analyzing samples from aquatic rather than terrestrial environments.

Table X. *Relative toxicities of pesticides to Daphnia pulex.*

Pesticide	48 hr LC$_{50}$ (ppm)
DDT	0.4
Diazinon	0.9
Malathion	2
Ethyl Guthion[a]	3
Carbaryl	6
Toxaphene	15
Endrin	20
Pyrethrins	25
Heptachlor	42
Dieldrin	250
Lindane	460
Trifluralin	240
Diuron	1,400
Sodium arsenite	1,800
Silvex	2,400
2,4-D	3,200
Dichlobenil	3,700
Fenac	4,500

[a] Azinphosethyl.

The so-called third generation insecticides, one class of which is insect growth regulators (IGR), were generated with the purpose of having greater desirability as pesticides with several requirements: the absence of undesirable effects on man, wildlife, and the environment and compatibility with modern insect pest management principles. Although the majority of new compounds in this class is the result of research on insect juvenile hormones, the general term IGR also accommodates compounds having a different mode of action (Staal 1975).

Methoprene [isopropyl (2E-4E)-11-methoxy-3,7,11-trimethyl-2,4-dodecadienoate (Altosid®)], the mosquito-specific IGR, was applied as a larvicide to irrigated pastures for the control of *Aedes nigromaculis* and found to be detrimental to aquatic Diptera (Chironomidae, Ephydridae, and Psycodidae). Similar applications to a pond had no effect on zoo-

plankton (Cladocera and Copepoda) (MIURA and TAKAHASHI 1973). STEEL-
MAN and SCHILLINGS (1972), however, observed that the number of
Dytiscidae beetle larvae was significantly reduced when concentrations
as low as 0.5 ppm were applied to a rice field as mosquito larvicides.

Farm ponds treated with diflubenzuron [(Dimilin®), 1-(4-chloro-
phenyl) 3-(2,6-difluorobenzoyl)urea] showed suppression of crustacean
zooplankton at all rates. Cladocerans were more susceptible than copepods and required longer recovery periods. Rotifer and algal populations
were not affected by the treatments (APPERSON et al. 1978).

More recently, ALI and MULLA (1978) studied the effects of diflubenzuron on planktonic, nektonic, and benthic invertebrates in a recrea-
tional lake in southern California. The spring application severely affected
cladoceran populations, and completely eliminated 3 species of copepods
within 1 wk. Recoveries required 11 wk to 6 mon, depending on species.
The prolonged absence of these nontarget organisms was thought to
be due to the seasonal cycles of the species and their inability to rapidly
recolonize the treated habitat. Nymphs of the mayfly Caenis sp. were
reduced 99%, recovering in 6 wk.

Pesticides do not always interact with aquatic microorganisms as
predicted. Generally all pesticides are toxic to all microorganisms at some
dosage. Toxicity as measured in reported studies includes changes in
growth rate, metabolic rate, and photosynthesis.

The phenylureas are the most toxic herbicides to phytoplankton,
while the cyclodienes are the most toxic insecticides (CAIN 1965). DDT
can reduce photosynthesis in phytoplankton, and is also the most toxic
material to many crustaceae.

Aquatic microorganisms absorb and concentrate pesticides from
water apparently inversely related to the water solubility of the compound. Living organisms do not seem to be any more efficient than dead
organisms in this seemingly nonspecific, physical property of microorganisms.

Metabolism of pesticides in microorganisms is as varied as in vertebrates. Any of the small forms will metabolize any of the pesticides to
some extent, perhaps with the exception of dieldrin. In the case of DDT
there are probably 2 routes of metabolism: aerobic, leading to the
formation of DDE, and anaerobic, which produces DDD. In the case
of actinomycetes metabolism occurred only during the active growth
phase, stopping completely when growth ceased. Phosphate insecticides
are readily metabolized by all bacteria, actinomycetes, and fungi and
algae examined. The 5 classes of herbicides are probably attacked by
all forms, some being highly selective (WARE and ROAN 1971).

This review has dealt essentially with primary effects of pesticides on
aquatic microorganisms and plankton. A longer range approach has
been taken in an excellent review by HURLBERT (1975) dealing with
secondary effects, those ecosystem changes that follow and result from
these primary effects.

IX. Plants
a) Insecticides

Certain insecticides have been known to cause typical phytotoxic effects on crop plants, e.g., tip burn, leaf burn, and stunting. However, certain organophosphate insecticides have had detrimental effects on cotton, some of which have not been explained and are frequently not reproducible (LEIGH 1977). Several wide-spread cotton stand failures were observed following treatment with the systemic insecticide phorate in 1959 to 1960. It was found that low temperatures and high soil moisture in combination with phorate seed treatment did indeed reduce germination substantially (HACSKAYLO and RANNEY 1961).

In North Carolina early season application of methyl parathion to cotton caused significant delays in fruiting and maturity as well as significant reductions in cotton yields in all experiments. Applications made prior to squaring reduced square production and early boll maturity up to 75%, resulting in yield reductions of as much as 30%. Monocrotophos on a full-season schedule caused maturity delays and yield reductions which were not significant, while malathion and dimethoate had no effect (BRADLEY and CORBIN 1974).

Soon after the introduction of systemic organophosphorus insecticides, WOLFENBARGER (1948) suggested that increased potato yields were in part a result of nutritional value gained by plants from the insecticides parathion and TEPP. HALL (1950) investigated the physiological effects produced by TEPP on carnation and tomato, and reported stimulatory effects at low application rates, and adverse growth and fruiting responses with higher application rates.

Translocation of some chlorinated cyclodiene insecticides from soils to above-ground parts, particularly oil seeds, has been reported from many areas. Notable among the identified plants are alfalfa and heptachlor and soybeans, peanuts, carrots and dieldrin (BECK et al. 1962, BRUCE et al. 1966, HARDEE et al. 1964, LICHTENSTEIN et al. 1964, LICHTENSTEIN and SCHULZ 1960, LICHTENSTEIN et al. 1965). DDT has never been considered a material which is actually translocated, though it does attach readily to roots, while BHC moves from soil into most plants (BRASS and WARE 1960, WARE 1968).

b) Herbicides

Aside from the bizarre effects of 2,4-D and other hormone-type herbicides on sensitive crops such as grapes and cotton, there are more subtle, long-range effects of herbicides on plants. The inactivation and dissipation rates of herbicides in soil vary widely. Some are inactive in the soil and have no residual phytotoxicity, while others exhibit various levels of phytotoxicity and different rates of dissipation from soils.

Residues of herbicides that persist from spring or summer to the next fall and spring frequently injure sensitive crops. Cereals planted in the

196 G. W. WARE

fall after crops that have received summer applications of atrazine, sima-
zine, diuron, diphenamid, and some related herbicides are most vulnerable. Injury to soybeans, sugar beets, oats, and forage grasses and
legumes has been encountered 10 to 12 mon after applications of atrazine
to corn. Tobacco, cotton, peanuts, and soybeans have been injured by
fenac residues 1 and 2 yr after application at rates used for selective
weed control in corn, and diphenamid has persisted for 10 to 12 mon
and injured oat seedlings.

Though low levels of phytotoxic residues have persisted from one
season to the next, data from many sources indicate that accumulation
of massive levels of selective herbicides is unlikely. Inherent phytotoxicity provides a natural indicator for residues and a defense against
accumulation of herbicides that are used for selective weed control in
crops (SHEETS and HARRIS 1965).

There has been an increase in the incidence of various plant diseases
caused by the application of herbicides, shown by greenhouse and field
studies over the past two decades. These include air-borne diseases
such as leaf diseases caused by *Alternaria solani* and *Erysiphe graminis*,
and soil-borne diseases such as seedling damping-off caused by *Rhizoctonia solani*, and vascular wild diseases caused by Fusarium organisms.
The wide range encountered indicates that the phenomenon of disease
increase due to herbicides is not restricted to a specific group of herbicides, pathogens, or crops. Four mechanisms have been suggested that
may be involved in disease increase: (1) direct stimulatory effects of
the herbicide on a pathogen, (2) effect of the herbicide on the virulence
of the pathogen, (3) effect on host susceptibility, and (4) effects on
relationships between pathogens and other organisms (KATAN and ESHEL
1973).

Summary

The effects of pesticides on nontarget organisms are not merely ob-
ject lessons of the past. All effects discussed, those that were omitted
from the review, and some that are yet unidentified are being produced
at the present. Their magnitude is likely not as great now as when most
attention was being devoted to the organochlorine decade (1960 to
1970). It was the organochlorine insecticides and their metabolites that
caused the most disruptive influences through time. Because their volume has declined an estimated 90% since the year of greatest use (1964),
their major nontarget influences may well be attenuated to a no-effect
level by the year 2000.

With our present pesticide use patterns, and the restraints surrounding these uses, the nontargets in greatest jeopardy today are obviously
those closest to application, the beneficial arthropods. Specifically, these
are the predators, parasites, and pollinating insects. It is not uncommon
to deplete populations of predators by 80% with the first application of

insecticide to a crop. From that moment on, throughout the growing season, the pests have the advantage and require repeated applications, each one depressing the natural enemy population more than the pests. Aquatic organisms are probably the second most jeopardized. From this review, it would appear that of the three classes of the most-used pesticides, the insecticides still offer the greatest potential for detrimental nontarget effects, and the fungicides the least. Herbicides, because of their varied modes of action and range of persistence, fall between the extremes. We may further conclude that soil microorganisms and arthropods and plants are the least endangered.

References

ADAMS, C. H., and W. H CROSS: Insecticide resistance in *Bracon mellitor*, a parasite of the boll weevil. J. Econ. Entomol. **60**, 1016 (1967).

ALEXANDER, M.: Microbial degradation of pesticides. In: F. Matsumura, G. M. Boush, and T. Misato (eds.): Environmental toxicology of pesticides, p. 365. New York: Academic Press (1972).

ALI, A., and M. S. MULLA: Impact of the insect growth regulator Diflubenzuron on invertebrates in a residential-recreational lake. Arch. Environ. Contam. Toxicol. **7**, 483 (1978).

ANONYMOUS: Residues in fatty tissues, brain, and milk of cattle from insecticides applied for grasshopper control on rangeland. J. Econ. Entomol. **52**, 1206 (1959).

APPERSON, C. S., C. H. SCHAEFER, A. S. COLWELL, G. H. WERNER, N. L. ANDERSON, E. F. DUPRAS, JR., and D. R. LOGANECKER: Effects of diflubenzuron on *Chaoborus astictopus* and nontarget organisms and persistence of diflubenzuron in lentic habitats. J. Econ. Entomol. **71**, 521 (1978).

ATKINS, E. L., L. D. ANDERSON, D. KELLUM, and K. W. NEUMAN: Protecting honey bees from pesticides. Div. Agr. Sci., Univ. Calif., Riverside. Leaflet 2883 (1977).

———, E. A. GREYWOOD, and R. L. MacDONALD: Toxicity of pesticides and other agricultural chemicals to honey bees. Univ. of Calif. Div. Agr. Sci. Leaflet 2287 (1973).

BARTHA, R., R. P. LANZILOTTA, and D. PRAMER: Stability and effects of some pesticides in soil. Applied Microbiol. **15**, 67 (1967).

BECK, E. W., L. H. DAWSEY, D. W. WOODHAM, D. B. LEUCK, and L. W. MORGAN: Insecticide residues on peanuts grown in soil treated with granular aldrin and heptachlor. J. Econ. Entomol. **55**, 953 (1962).

BERRY, J. H.: Pesticides and energy utilization. Paper presented AAAS Ann. Meeting, Washington, D. C., 17 Feb. (1978).

BEVENUE, A.: The bioconcentration aspects of DDT in the environment. Residue Reviews **61**, 37 (1976).

BOLLAG, J. M.: Biochemical transformation of pesticides by soil fungi. Crit. Rev. Microbiol. **2**, 35 (1972).

BOTTRELL, D. G., and D. R. RUMMEL: Response of Heliothis populations to insecticides applied in an area-wide reproduction diapause boll weevil suppression program. J. Econ. Entomol. **71**, 87 (1978).

BRADLEY, J. R., and F. T. CORBIN: Effects of organophosphate insecticides, especially methyl parathion, on fruiting, maturity, and yield of cotton. Proc. Beltwide Cotton Prod. Res. Conf. Dallas, TX. Jan. 7–9 (1974).

BRASS, C. L., and G. W. WARE: BHC translocation from treated soil and the effect on growth of red clover. J. Econ. Entomol. **53**, 110 (1960).

BROOKS, G. T.: Chlorinated insecticides. II. Biological and environmental aspects. Cleveland, Ohio: CRC Press (1974).

BROWN, A. W. A.: Ecology of pesticides. New York: Wiley (1978).

198 G. W. WARE

——, and R. PAL: Insecticide resistance in arthropods. Monogr. Ser. No. 38. Rome: World Health Organization (1971).

BRUCE, W. N., G. C. DECKER, and J. G. WILSON: The relationship of the levels of insecticide contamination of crop seeds to their fat content and soil concentration of aldrin, heptachlor, and their epoxides. J. Econ. Entomol. 59, 179 (1966).

——, R. P. LINK, and G. C. DECKER: Storage of heptachlor epoxide in the body fat and its excretion in milk of dairy cows fed heptachlor in their diets. J. Agr. Food Chem. 13, 63 (1965).

CAIN, S. A.: Pesticides in the environment, with special attention to aquatic biology resources. Rep. on U.S.-Japan Planning Meet. Pest. Res., Honolulu, p. 12 (1965).

CAMPBELL, J. L., L. A. RICHARDSON, and M. L. SCHAFER: Insecticide residues in the human diet. Arch. Environ. Health 10, 831 (1965).

CARSON, R. L.: Silent spring. Boston: Houghton Mifflin Co. (1962).

COPE, O. B.: Interactions between pesticides and wildlife. Ann. Rev. Entomol. 16, 325 (1971).

—— Contamination of the freshwater ecosystem by pesticides. J. Applied Ecol. 3 (Suppl.), 33 (1966).

DEBACH, P.: Cottony-cushion scale, vedalia and DDT in central California. Citrograph 32, 406 (1947).

——, and B. R. BARTLETT: Effects of insecticides on biological control of insect pests of citrus. J. Econ. Entomol. 44, 372 (1951).

DRAKE, J. L., G. W. WARE, and F. G. WERNER: Insecticidal effects on soil arthropods. J. Econ. Entomol. 64, 842 (1971).

DUGGAN, R. E., and J. R. WEATHERWAX: Dietary intake of pesticide chemicals. Science 157, 1006 (1967).

EDWARDS, C. A.: Pesticide residues in soil and water. In C. A. Edwards (ed.): Environmental pollution by pesticides, p. 409. New York: Plenum Press (1973).

——, and A. R. THOMPSON: Pesticides and the soil fauna. Residue Reviews 45, 1 (1973).

——, E. B. DENNIS, and D. W. EMPSON: Pesticides and the soil fauna: Effects of aldrin and DDT in an arable field. Ann. Applied Biol. 60, 11 (1967).

FABACHER, D. L., J. D. DAVIS, and D. A. FABACHER: Apparent potentiation of the cotton defoliant DEF® by methyl parathion in mosquitofish. Bull. Environ. Contam. Toxicol. 16, 716 (1976).

FLINT, M. L., and R. VAN DEN BOSCH: A source book on integrated pest management. Intl. Center Integrated & Biol. Control. Berkeley, Univ. Calif. (1977).

FRAWLEY, J. P., H. N. FUYAT, E. C. HAGEN, J. R. BLAKE, and O. G. FITZHUGH: Marked potentiation in mammalian toxicity from simultaneous administration of two anticholinesterase compounds. J. Pharm. Exptl. Therap. 121, 96 (1957).

GUNTHER, F. A., W. E. WESTLAKE, and P. S. JAGLAN: Reported solubilities of 738 pesticide chemicals in water. Residue Reviews 20, 1 (1968).

HACSKAYLO, J., and C. D. RANNEY: Emergence of phorate-treated cotton seed as affected by substrate moisture and temperature. J. Econ. Entomol. 54, 296 (1961).

HALL, W. C.: Morphological and physiological responses of carnation and tomato to organic phosphorus insecticides and inorganic soil phosphorus. Plant Physiol. 26, 502 (1950).

HARDEE, D. D., W. H. GUTENMANN, G. I. KEENAN, G. G. GYRISCO, D. J. LISK, W. H. FOX, G. W. TRIMBEYER, and R. F. HOLLAND: Residues of heptachlor epoxide and telodrin in milk from cows fed at part per billion insecticide levels. J. Econ. Entomol. 56, 404 (1964).

HELLING, C. S., P. C. KEARNEY, and M. ALEXANDER: Behavior of pesticides in soils. Adv. Agron. 23, 147 (1971).

HICKEY, J. J., and D. W. ANDERSON: Chlorinated hydrocrabons and eggshell changes in raptorial fish-eating birds. Science 162, 271 (1968).

——, J. A. KEITH, and F. B. COON: An exploration of pesticides in a Lake Michigan estuary. J. Applied Ecol. 3 (Suppl.), 141 (1966).

HUBER, R. T.: Private communication (1979).
HUNT, E. G., and A. I. BISHOFF: Inimical effects on wildlife of periodic DDD applications to Clear Lake. Calif. Fish and Game 46, 10 (1960).
HURLBERT, S. H.: Secondary effects of pesticides on aquatic ecosystems. Residue Reviews 57, 81 (1975).
ICAITI, R. B.: An environmental and economic study of the consequences of pesticide use in Central American cotton production. Final Report (Phase 1), Guatemala (1976).
JOHNSON, B. T., C. R. SAUNDERS, H. O. SANDERS, and R. S. CAMPBELL: Biological magnification and degradation of DDT and aldrin by freshwater invertebrates. J. Fisheries Res. Bd., Canada 28, 705 (1971).
JOHNSEN, R. E.: DDT metabolism in microbial systems. Residue Reviews 61, 1 (1976).
KATAN, J., and Y. ESHEL: Interactions between herbicides and plant pathogens. Residue Reviews 45, 145 (1973).
KEENLEYSIDE, M. H. A.: Effects of spruce budworm control on salmon and other fishes in New Brunswick. Can. Fish Culturist 24, 17 (1959).
KEEVER, D. W., J. R. BRADLEY, JR., and M. C. GANYARD: Effects of diflubenzuron (Dimilin) on selected beneficial arthropods in cotton fields. Environ. Entomol. 6, 732 (1977).
KUHR, R. J., and H. W. DOROUGH: Carbamate insecticides: Chemistry, biochemistry, and toxicology. Cleveland, Ohio: CRC Press (1976).
LAVEGLIA, J., and P. A. DAHM: Degradation of organophosphorus and carbamate insecticides in the soil and by soil microorganisms. Ann. Rev. Entomol. 22, 483 (1977).
LEIGH, T. F.: Private communication (1979).
—— Insecticides and cotton yields. Proc. Beltwide Cotton Prod. Res. Conf. Atlanta, GA Jan. 10–13 (1977).
LEVIN, M. D., W. B. FORSYTH, G. L. FAIRBROTHER, and F. B. SKINNER: Impact on colonies of honey bees of ultra-low-volume (undiluted) malathion applied for control of grasshoppers. J. Econ. Entomol. 61, 48 (1968).
LIANG, T. T., and E. P. LICHTENSTEIN: Synergism of insecticides by herbicides: Effect of environmental factors. Science 186, 1128 (1974).
LICHTENSTEIN, E. P., and K. R. SCHULZ: Translocation of some chlorinated hydrocarbon insecticides into the aerial parts of pea plants. J. Agr. Food Chem. 8, 452 (1960).
——, G. R. MYRDAL, and K. R. SCHULZ: Effect of formulation and mode of application on the loss of aldrin and its epoxide from soils and their translocation into carrots. J. Econ. Entomol. 57, 133 (1964).
——, K. R. SCHULZ, R. F. SKRENTING, and P. A. STITT: Insecticidal residues in cucumbers and alfalfa grown on aldrin or heptachlor treated soils. J. Econ. Entomol. 58, 742 (1965).
LUCKENS, M. M., and W. H. DAVIS: Bats: Sensitivity to DDT. Science 146, 948 (1964).
MACEK, K. J.: Screening of pesticides against fish. In: Progress in sport fishery research, 1968. Bur. Sport Fish. Wildl., U.S. Res. Pub. 77, p. 92 (1969).
MATSUMURA, F., G. M. BOUSH, and T. MISATO (eds.): Environmental toxicology of pesticides. New York: Academic Press (1972).
MIURA, T., and R. M. TAKAHASHI: Insect development inhibitors. 3. Effects on nontarget aquatic organisms. J. Econ. Entomol. 66, 917 (1973).
MOFFETT, J. O., R. H. MACDONALD, and M. D. LEVIN: Toxicity of carbaryl-contaminated pollen to adult bees. J. Econ. Entomol. 63, 475 (1970).
NEWSOM, L. D.: Consequences of insecticide use on nontarget organisms. Ann. Rev. Entomol. 12, 257 (1967).
PAPE, D. J.: Unpublished data (1979).
PEAKALL, D. B.: Pesticide-induced enzyme breakdown of steroids in birds. Nature 216, 505 (1967).

200   G. W. WARE

PFISTER, R. M.: Interactions of halogenated pesticides and microorganisms: a review. Crit. Rev. Microbiol. 2, 1 (1972).
PIMENTEL, D.: Ecological effects of pesticides on non-target species. Executive office of the President, Office of Science and Technology, Supt. of Documents. U.S. Govt. Printing Office, Stock No. 4106-0029, Washington, D. C. (1971).
PLAPP, F. W., W. S. BAGLEY, G. A. CHAPMAN, and G. W. EDDY: Synergism of malathion against resistant house flies and mosquitoes. J. Econ. Entomol. 56, 643 (1963).
POLIVKA, J. B.: Effects of insecticides upon earthworm populations. Ohio J. Sci. 51, 195 (1951).
—— More about the effect of insecticides on earthworm populations. Presented 62nd Ann. Meet. Ohio Acad. Sci. Columbus. Unpublished (1953).
RAW, F.: The accumulation of residues of chlorinated hydrocarbon insecticides in soil invertebrates. Rep. Rothamst. Exp. Sta. p. 149 (1963).
RIDGWAY, R. L., P. D. LINGREN, C. B. COWAN, JR., and J. W. DAVIS: Populations of arthropod predators and Heliothis spp. after applications of systemic insecticides to cotton. J. Econ Entomol. 60, 1012 (1967).
SCHECHTER, M. S.: Private communication (1969).
SCHWARTZ, H. G., JR.: Microbial degradation of pesticides in aqueous solutions. J. Water Pollution Control Fed. 39, 1701 (1967).
SETHUNATHAN, N.: Microbial degradation of insecticides in flooded soil and in anaerobic cultures. Residue Reviews 47, 143 (1973).
SEUME, F. W., and R. D. O'BRIEN: Potentiation of the toxicity to insects and mice of phosphorothionates containing carboxyester and carboxyamide groups. Toxicol. Applied Pharmacol. 2, 495 (1960).
SHEETS, T. J., and C. I. HARRIS: Herbicide residues in soils and their phytotoxicities to crops grown in rotations. Residue Reviews 11, 119 (1965).
SHOUR, M. H., and L. A. CROWDER: Unpublished data. (1979).
STAAL, G. B.: Insect growth regulators with juvenile hormone activity. Ann. Rev. Entomol. 20, 417 (1975).
STEELMAN, C. D., and P. E. SCHILLINGS: Effect of a juvenile hormone mimic on Psorophora confinnis and non-target aquatic insects. Mosq. News 32, 350 (1972).
STICKEL, L. F.: Pesticide residues in birds and mammals. In C. A. Edwards (ed.): Environmental pollution by pesticides, p. 254. New York: Plenum Press (1973).
——, W. H. STICKEL, and R. CHRISTENSEN: Residues of DDT in brains and body of birds that died on dosage and in survivors. Science 151, 1549 (1966).
STICKEL, W. H., L. F. STICKEL, and J. W. SPANN: Tissue residues of dieldrin in relation to mortality in birds and mammals. In: Chemical fallout: Current research on persistent pesticides, p. 174. Proc. 1st Rochester Conf. on Toxicity. Springfield, Ill.: Charles C. Thomas (1969).
STREET, J. C., F. L. MAYER, and D. J. WAGSTAFF: Ecological significance of pesticide interactions. In W. B. Deickmann (ed.): Pesticides symposia. Miami, FL: Halos and Associates (1970).
TU, C. M., and J. R. W. MILES: Interactions between insecticides and soil microbes. Residue Reviews 64, 19 (1976).
UKELES, R.: Growth of pure cultures of marine phytoplankton in the presence of toxicants. Applied Microbiol. 10, 532 (1962).
U. S. Department of Agriculture: Beekeeper indemnity payment program. ASCS Handbook Short Ref. 4-LD (1971).
WARE, G. W.: DDT-¹⁴C translocation in alfalfa. J. Econ. Entomol. 61, 1451 (1968).
—— Bees in agriculture—Their importance and problems. In: The indispensable honeybee. Report Beekeeping Ind. Conf., Beltsville, MD, Feb. 12–13 (1973).
—— Ecological history of DDT in Arizona. J. Az. Acad. Sci. 9, 61 (1974).
—— Effects of DDT on reproduction in higher animals. Residue Reviews 59, 119 (1975).
—— Statement read at the hearing of the Board of Pesticide Control regarding the

banning of toxaphene in agriculture. Dept. Entomol., Univ. Arizona, Tucson (1979).

——, and C. C. ROAN: Interaction of pesticides with aquatic microorganisms and plankton. Residue Reviews 33, 15 (1971).

——, and E. C. NABER: Lindane in eggs and chicken tissues. J. Econ. Entomol. 54, 675 (1961).

WATSON, T. F.: Private communication (1979).

——, L. MOORE, and G. W. WARE: Practical insect pest management. San Francisco, CA: W. H. Freeman & Co. (1976).

WEISS, C. M.: Response of fish to sub-lethal exposures of organic insecticides. Sewage Ind. Wastes 31, 580 (1959).

WESTLAKE, W. E., and F. A. GUNTHER: Organic pesticides in the environment. Adv. Chem. Series 60, 110 (1966).

WOLFENBARGER, D. O.: Nutritional value of phosphatic insecticides. J. Econ. Entomol. 41, 818 (1948).

WOODWELL, G. M.: Toxic substances and ecological cycles. Sci. Amer. 216, 24 (1967).

WRIGHT, M. A.: Effects of benomyl and some other systemic fungicides on earthworms. Ann. Applied Biol. 87, 520 (1977).

Manuscript received April 27, 1979; accepted January 24, 1980.

Pesticide residue analysis of water and sediments: Potential problems and some philosophy

By

John W. Hylin*

I. Introduction

The analysis of water and sediments for residues of pesticides has occupied many research workers throughout the world. This present paper will not attempt to review this voluminous amount of work, but will instead discuss the underlying theoretical and philosophical concepts on how-to and how-not-to analyze water and sediments for pesticide residues. The first point that must be clarified is what is meant by "water." There are many liquids that are referred to as water: drinking water, rainwater, fresh water, salt water, sea water, and even some beer and rice wine. All of the above are impure forms of water. As I will use the term in this paper, "water" is pure H_2O, a substance that is difficult to prepare in the pure state. We think of distilled water as "pure" water but studies have shown that such need not be the case (BEVENUE et al. 1971 b). When we discuss the analysis of pesticide residues in water it is important to remember to define the quality of water under study since this markedly affects the approach to problem solution.

_____

* Agricultural Biochemistry, University of Hawaii, Honolulu, Hawaii, 96822. Published with the approval of the Hawaii Institute of Tropical Agriculture and Human Resources as Technical Paper 2485. Presented in part at the April 1979 US-ROC Cooperative Science Program seminar on "Environmental Problems Associated with Pesticide Usage in the Intensive Agricultural System," Taipei, Taiwan, Republic of China, as sponsored by the National Science Foundation (U.S.A.) and the National Science Council (R.O.C.).

204   J. W. HYLIN

For instance, when one reads of pesticide residues in water one
should assume that the residues are dissolved in the water, that is, that
the solute and the solvent form a homogenous phase. However, we
know that most pesticides in use today are not very soluble in water,
and some are so poorly soluble that only the purist would refrain from
describing these compounds as insoluble in water (GUNTHER et al. 1968).
The organochlorine pesticides, that have already occupied our attention
in this seminar, are so water-insoluble that it becomes very difficult to
determine their true water solubility. The least soluble is probably DDT
which has a presumed solubility in water of 1 μg/L or 1 ppb. How then
do we account for reports in the literature of finding 50 or 100 ppb of
DDT in water? The explanation for these results is at least two-fold.
First, the water analyzed was not pure and contained substantial quantities of other solutes and/or particulates. Secondly, the DDT was not
in solution but was absorbed on the particulates or was present itself
as particulates or in colloidal suspension.

Now a comment to the above might be "so what?" The DDT is present and it is measurable at the levels indicated, so it is *in* the water.
From the point of a nonscientist this is probably an acceptable statement. However, the fact that DDT is not in true solution has marked
effects on its biological properties and effects. Particulates do not usually
pass through cell membranes so although substantial amounts of a pesticide residue may be present in the water in the form of particulates,
these residues may have no biological effects. KOHN (1980) has mentioned the case of the herbicide paraquat absorbed on monmorillonite
clay. Paraquat is very water-soluble so with pure water one could have
substantial concentrations of paraquat that would be biologically active.
However, in the presence of even a small amount of suspended clay
particulates, the same quantity of paraquat is not biologically active.
This is presumably because the paraquat is so firmly bound to the clay
particles as to be biologically unavailable.

Although the capacity of some clay minerals to bind organic molecules is enormous, this binding capacity is nonetheless finite. An example
of this phenomenon is the appearance of what are believed to be symptoms of phytotoxicity in papaya due to excessive paraquat residues in
the soil. One suggested explanation is that the binding sites in this soil
are almost saturated with paraquat and an equilibrium has been established between bound paraquat and free paraquat in soil water.

Therefore, the very occurrence of a pesticide residue in water is
dependent upon at least two factors: the true solubility of the pesticide
residue in pure water and the purity of the water being studied.

II. Water purity

The question of water purity will now be discussed, but first some-
thing should be said about the value of performing analyses at the micro-

gram and picogram levels. The toxicological significance of such small quantities is difficult to assess even for very toxic substances such as tetrachlorodibenzodioxin and the aflatoxins. Nevertheless, when dealing with potable water even such small quantities of residues may have adverse effects since we consume several liters of water/day in one form or another. Furthermore, most of the food we eat comes in contract with water prior to consumption. This may occur when prepared for market, in the market itself, in food processing, and in the final preparation for eating, to name a few possibilities. If in any of these instances the water contains dissolved or suspended residues these may be transferred to the food. Sometimes the pesticide residues may be more soluble in the components of the food, or may be preferentially absorbed onto the food surface during washing, marketing, or cooking.

These points have not been lost on regulatory agencies such that some have proposed water quality standards with maximum levels of pollutants stated in nanogram and picogram/L levels, i.e., ppb and ppt (Anonymous 1968). Most analytical procedures for pesticide residues involve water at some stage, if only to rinse the glassware after washing. When trying to analyze for residues at the nanogram or picogram levels, it is therefore essential that all the materials that come in contact with the analytical sample, including water, be as pure as possible when nonspecific detectors, such as electron capture, are used (Bevenue et al. 1971 a).

One possible solution to this problem is to use more specific detectors in gas chromatography. However, experience has shown that even with these detectors the analyst obtains erroneous data due to impurities in the water used in the analytical procedure.

To avoid spurious peaks or high baseline noise the water used for nanogram and picogram residue analysis must be of the highest quality obtainable and recently prepared. Conventional stills produce water that is generally not suitable for ultratrace analysis of organics by electron-capture gas chromatography (Bevenue et al. 1971 b). Ion exchangers produce even worse water. In our experience the only truly suitable water for ultratrace analyses is by distillation from an all-glass or quartz apparatus with collection in an all-glass vessel. And by all-glass I mean all glass: no Teflon stopcocks, no grease on ground joints, and every opening protected by an acid, alkali, and particulate trap. The water to be distilled should be treated with potassium permanganate and distillation should occur from a solution containing permanganate in an apparatus similar to that described by Ballentine (1954). The distillate should be discarded after one month. Naturally, the quality of the distillate should be checked by extracting a subsample with a suitable solvent and examination of the extract by electron-capture gas chromatography.

Glassware used for the collection of water samples for nanogram and picogram analysis, and such glassware as is used in the analytical procedure, should be cleaned with hot chromic acid, rinsed successively with pure water, nanograde hexane, nanograde acetone, and dried at 300°C for 16 hr if the analyst is to minimize noise and spurious peaks in the chromatogram. Glassware used for the analysis of pesticide residues in water at the nanogram and picogram levels should be kept separate from other laboratory glassware. The cleaning procedures described above were developed, out of necessity, for a project to analyze organochlorine residues in water, sediment, algae, and fish from around the Hawaiian Islands in 1970 to 1971 (BEVENUE et al. 1972).

III. Sampling and sample preparation

The quality of analytical results is almost completely dependent on the quality of the sample, and this is nowhere more evident than in the analysis of water for organochlorine residues by electron-capture gas chromatography. Although this analytical tool is very sensitive for the detection of residues of organochlorine pesticides, it is relatively nonspecific. Hence it is necessary to maintain the integrity of the analytical sample by using pure substances, ultra-clean glassware, and innumerable control analyses. Volumes of water that can be conveniently handled are usually restricted to about 4 L. Thus, occasional or gram samples may be collected in 1-gal bottles that have been specially cleaned. If the total residue in the raw water is of interest, extraction of this residue with a suitable solvent can take place in the bottle, thus minimizing the number of surfaces contacted by the sample. If the partition of residues between the particulates in the water and the water itself is of interest, then some means of segregating the two phases must be used. Particulates down to the micron size may be collected by filtration of the sample through special filtration media (Millipore, Nucleopore, etc.) or by sedimentation in a high-speed centrifuge. The absence of interferring contaminants in the filter media and the centrifuge containers must be verified if nanogram and picogram quantities of residue are being sought. These segregation procedures are time-consuming, and cleaning the materials and glassware to eliminate contamination may be very difficult.

We have found filtration through glass fiber filter paper to be a satisfactory alternative if the ultimate in sensitivity is not required. Glass fiber filter paper can be cleaned by the procedures described for glassware, is quite inert, and permits rapid filtration.

Following separation, the water and particulates are extracted with suitable solvents and analyzed separately. Estimation of the quantity of particulates is particularly easy with glass fiber paper since it can be dried to constant weight in a desiccator at room temperature under conditions which will minimize transformation of residues adsorbed to any collected particulates. Extraction of the dried filter paper can be done with or without rehydration as the analyst prefers.

In some cases it is necessary or desirable to try to detect residues at sensitivities greater than permited by a 4-L sample. Although continuous solvent extraction in complicated and expensive glass extractors have been used in the past, the more commonly used techniques now involve the passage of a measured volume of water through a column containing a selective adsorbent. The adsorbent may be a polymeric resin such as the *Rohm & Haas* XAD series, an ion exchanger, polyurethane foam, or even a specially prepared version of a gas chromatographic column stationary phase. Activated charcoal was originally recommended for this procedure but was found to give rise to so many artifacts that its use is not now recommended. Although substantial concentration of residues can be achieved by this technique, great care must be taken to prepare the adsorbents so as to minimize background and artifacts. The polymeric XAD resins are usually exhaustively extracted with methanol before use. Ion-exchange resins must be extracted with solvent and then cycled several times so as to minimize background interferences, and the gas chromatographic liquid phase absorbents are best prepared by the procedure of Aue *et al.* (1973). Although these techniques permit processing large volumes of water, care must be used in interpreting the significance of residue values found by these specialized methods.

IV. Sediments

Sediment is usually defined as particulate matter that has settled out of water onto the stream or riverbed or lake or ocean floor. It usually consists of a complex mixture of inorganic minerals and organic substances in a dynamic and changing condition and in some sort of equilibrium with the overlying water. Sediment particles may be resuspended and moved to another site as the result of changes in the volume and velocity of water passing a given site. Thus the composition of sediments collected at any geographical point will vary over time. This is particularly true for the pesticide residue component of the sediment which may be bound to inorganic minerals or organic matter. Just because one finds a reduction in residue levels over a period of time at one site does not necessarily mean that degradation or metabolism of the residues has taken place at that site. The only valid conclusion that can be drawn on such isolated data is that the residue levels at that site have changed. A thorough knowledge of water movement in a system and adequate sampling may permit more accurate prediction of pesticide residue behavior in sediments associated with a given site. In a study of the "Quality of Coastal Waters in Hawaii," one site consisted of a brackish water lagoon connected at only one point with a bay which then merged with the open ocean. Residues of persistent organochlorine pesticides used for subterranean termite control entered the lagoon on particulate matter carried by run-off water during rainstorms. By judicious selection of sampling sites and times it was possible to describe the movement

of sediments and their associated pesticide residues from the point of
introduction into the system until their dispersal in the open ocean
(LAU 1972).

The complexity of sediment samples prevents analyses of pesticide
residues below the mg/kg (ppm) level except in special cases. In addition, in most ecosystems, sediment is the ultimate sink for residues so
the quantities found are usually larger than the quantities found in
water. However, the exact distribution of residues between the aqueous
and sediment phases depends on the physicochemical characteristics of
the particular pesticide residue. As mentioned earlier, the concentration
of the components of a sediment-water biphase tend to approach equilibrium. Thus, poorly soluble substances such as the organochlorines
will be concentrated in the nonaqueous phase, the sediment. Very water-
soluble substances such as oxamyl will be found almost exclusively in
the water phase. Paraquat, although very water soluble, also strongly
binds to certain clay minerals so that residues of it will occur in sediment. Amphiphobic substances, such as pentachlorophenol, may be present predominantly in water or the sediment, depending on the pH.

V. Conclusions

Water and sediment are only two components of most ecosystems.
In order to understand the significance of pesticide residues in water
and sediment it is necessary to look at the biota associated with these
components of the complete ecosystem. An example of the type of distribution of pesticide residues among the components of an ecosystem
is illustrated in Table I. The increasing amount of residues found in
the larger organisms has been called biomagnification. However, in the
case of organochlorine pesticides, the larger amounts of residues found
in the larger animals are better correlated with higher fat contents since
these also increase with size of organism. A study designed to evaluate
the presence, amount, the significance of residues of any substance in

Table I. *Chlorinated pesticide residues in a canal environment*
(SHULTZ et al. 1975).

Sample specimen	Residues (ppm)			
	DDE	DDD	DDT	Dieldrin
Water	2×10^{-7}	2×10^{-6}	2×10^{-6}	1×10^{-5}
Algae	0.01	0.03	0.04	0.04
Sediment	0.04	0.12	0.07	0.04
Small fish	0.06	0.14	0.15	0.23
Elops hawaiensis (a carnivore)	0.18	0.57	0.10	0.14
Chanos chanos (a detrital feeder)	0.30	0.15	0.16	0.49

an environment, whether that substance is natural or man-made, organic or inorganic, must be planned with great attention to detail if erroneous data, and consequently erroneous conclusions, are to be avoided. The samples taken, the integrity of the samples prior to analysis, the reagents and glassware, the analytical procedure, and the interpretation of results are all areas where carelessness will jeopardize the whole study. The conclusion is therefore obvious: careful planning and execution of a residue study will result in reliable data. Any effort less than this is a waste of time.

Summary

The factors which must be considered in examining water and sediment for pesticide residues start with a decision as to which components will be separated and examined independently. Since most pesticide residues are not particularly soluble in water, they are found bound to particles in natural waters. Whether the gross sample will be examined for residues or whether the particles will be separated and examined separately from water are decisions the analyst must make during the planning stages of the study. All the components of the analytical stream, including sample collection equipment, must be such as to not introduce interfering contaminants into the analysis. Interferences may be minimized by careful cleaning and use of ultrapure reagents, or by selection of specific determinative steps insensitive to the impurities present. Analysis of sediments for residues is more difficult since sediments are even more heterogenous than natural water, including spatial and temporal variability. Usually residues concentrate in sediments so the sensitivity of the analytical procedure for them is less important than in the analysis of water. Finally, the other components of the water sediment ecosystem should be included in the study so that the significance of detectable residues may be assessed.

References

ANONYMOUS: Report of the committee on water quality criteria, p. 20. Federal Water Pollution Control Administration, Washington, D. C. (1968).

AUE, W. A., C. R. HASTINGS, and S. KAPILA: On the unexpected behavior of a common gas chromatographic phase. J. Chromatog. 77, 299 (1973).

BALLENTINE, R.: High efficiency still for pure water. Anal. Chem. 26, 549 (1954).

BEVENUE, A., J. W. HYLIN, Y. KAWANO, and T. W. KELLEY: Organochlorine pesticide residues in water, sediment, algae, and fish: Hawaii—1970–71. Pest. Monit. J. 6, 56 (1972).

——, T. W. KELLEY, and J. W. HYLIN: Problems in water analysis for pesticide residues. J. Chromatog. 54, 71 (1971 a).

——, J. N. OGATA, Y. KAWANO, and J. W. HYLIN: Potential problems with the use of distilled water in pesticide residue analysis. J. Chromatog. 60, 45 (1971 b).

GUNTHER, F. A., W. E. WESTLAKE, and P. S. JAGLAN: Reported solubilities of 738 pesticide chemicals in water. Residue Reviews 20, 1 (1968).

KOHN, G. K.: Bioassay as a monitoring tool. Residue Reviews **75**, 00 (1980).
LAU, L. S.: Second annual report, "Quality of Coastal Waters Project," 1972; NOAA
   Sea Grant, Water Resources Research Center, Univ. of Hawaii, Honolulu (1972).
SCHULTZ, C. D., W. L. YAUGER, and A. BEVENUE: Pesticides in fish from a Hawaiian
   canal. Bull. Environ. Contam. Toxicol. **13**, 206 (1975).

Manuscript received January 28, 1980; accepted January 28, 1980.

Subject Index

Subject Index

INFORMATION FOR AUTHORS

RESIDUE REVIEWS

(A BOOK SERIES CONCERNED WITH RESIDUES OF
PESTICIDES AND OTHER CONTAMINANTS
IN THE TOTAL ENVIRONMENT)

Edited by

Francis A. Gunther

Published by
Springer-Verlag New York • Heidelberg • Berlin

The original (ribbon) copy and one good xerox or other copy of the manuscript, complete with figures and tables, are required. Manuscripts will normally be published in the order in which they are received, reviewed, and accepted. They should be sent to the editor:

Professor Francis A. Gunther
Department of Entomology
University of California
Riverside, California 92502
Telephone: (714) 787-5804/5810 (office)
(714) 688-6666 (home)

1. Manuscript

The manuscript, in English, should be typewritten, double-spaced throughout, on one side of 8½ x 11 inch blank white paper, with at least one-inch margins. The first page of the manuscript should start with the title of the manuscript, name(s) of author(s), with author affiliation(s) as first-page starred footnotes, and "Contents" section. Pages should be numbered consecutively in arabic numerals, including those bearing figures and tables only. In titles, in-text outline headings and subheadings, figure legends, and table headings only the initial word, proper names, and universally capitalized words should be capitalized.

Footnotes should be inserted in text and numbered consecutively in the text using arabic numerals.

Tables should be typed on separate sheets and numbered consecutively within the text in roman numerals; they should bear a descriptive heading, in lower case, which is underscored with one line and which starts after the word "Table" and the appropriate roman numeral; *footnotes in tables* should be designated consecutively within a table by the lower-case alphabet. *Figures* (including photographs, graphs, and line drawings) should be numbered consecutively within the text in arabic numerals; each figure should be affixed to a separate page bearing a legend (below the figure) in lower case starting with the term "Fig." and a number.

2. Summary

A concise but informative summary (double-spaced) must conclude the text of each manuscript; it should summarize the significant content and major conclusions presented. It must not be longer than two 8½ × 11 inch pages of double-spaced typing. As a summary, it should be more informative than the usual abstract.

3. References

All papers, books, and other work cited in the text must be included in a "References" section (also double-spaced) at the end of the manuscript: If comprehensive papers on the same subject have been published, they should be cited but only for exceptional reasons should the bibliographic citations extend farther back than to these papers.

The references used *in the text* should consist of the COMPLETELY CAPITALIZED author's or authors' last name(s) where one or two authors are concerned; should there be more than two authors, only the first is named and *"et al."* is added. The publication year in parentheses should follow the name. If more than one paper by one author published in the same year is cited, the letters a, b, c, etc., should follow the year, e.g., "MEIER (1958 a) found . . .", or "This method is nonspecific (MEIER 1958 a)."

In the References section, the papers cited should appear in alphabetical order according to the last name of the first author; if more than one paper by an author or authors published in the same year is cited, the papers should be listed according to the year of publication followed by a, b, c, etc., as necessary. Papers published in periodicals should be cited with COMPLETELY CAPITALIZED names and initials of all authors, together with the *full title of the paper and preferably in its original language,* title of the periodical (abbreviated in accordance with *Chemical Abstracts'* "List of Periodicals Abstracted"), number of the volume (wavy underlined), initial page, and the year in parentheses. References to unpublished papers that have been submitted for publication should be cited in the same manner as other papers except the abbreviated journal name is followed by the words "In press" or "Accepted for publication" and the year in parentheses; personal communications are to be cited similarly.

In text and in the References section, citation of *governmental agencies, educational and research institutions and foundations, professional associations, and industrial companies* should consist of the full name as used by the organization completely underscored with one line and with initial capital letters only, followed by the appropriate reference information as specified above.

Examples:

EDWARDS, C. A., and E. B. DENNIS: Some effects of aldrin and DDT on the soil fauna of arable land. Nature 188, 767(1960).

GUNTHER, F. A., J. H. BARKLEY, and W. E. WESTLAKE: Worker environment research. II. Sampling and processing techniques for determining dislodgable pesticide residues on leaf surfaces. Bull. Environ. Contam. Toxicol. Accepted for publication (1974).

HESSLER, W.: Eine einfache Nachweismethode für Paraffin in Wachsgemischen. II. Mitt. Fette, Seifen, Anstrichmittel 58, 602(1956).

MELZER, H.: The qualitative and quantitative colorimetric determination of captan. Nachrbl. deut. Pflanzenschutzdienst 14, 193(1960).

Shell Chemical Co.: Letter to EPA's "Hazardous Materials Advisory Committee," Oct. 28(1971).

U.S. Environmental Protection Agency: Proposed toxicology guidelines. Fed. Register 37(183), 19383(1972).

Books should be cited with COMPLETELY CAPITALIZED name(s) and initials of the author(s), full title, edition or volume, page number(s), place of publication, publisher, and year of publication in parentheses.

Examples:

BEVENUE, A.: Gas chromatography. In G. Zweig (ed.): Analytical methods for pesticides, plant growth regulators, and food additives. Vol. I, p. 189. New York: Academic Press (1963).

DORMAL, S., and G. THOMAS: Répertoire toxicologique des pesticides, p. 48. Gembloux: J. Duculot (1960).

HARTE, C.: Physiologie der Organbildung, Genetik der Samenpflanzen. In:

Fortschritte der Botanik. Vol. 22, p. 315. Berlin-Göttingen-Heidelberg: Springer (1960).
METCALF, R. L.: Organic insecticides, their chemistry and mode of action. 2 ed., p. 51. New York-London: Interscience (1961).

4. Illustrations

Illustrations of any kind may be included only when indispensable for the comprehension of text; they should not be used in place of concise, clear explanations in text. Schematic line drawings must be drawn carefully and clearly. For other illustrations, clearly defined black-and-white glossy photographic prints are required. Should precisely placed indication darts (arrows) or letters be required on a photograph or other type of illustration, they should be marked neatly with a soft pencil on a duplicate copy or on an overlay, with the end of each dart (arrow) indicated by a fine pinprick; darts and lettering will be transferred to the illustrations by the publisher.

Photographs should be not less than five × seven inches in size. Unimportant and indistinct strips or areas on the edges of photographs should be marked on the back of the glossy print (pattern) with pencilled down-strokes, in order that the reproduction surface will not be unnecessarily large; alterations of photographs in galley-proof stage are not permitted. *Each photograph or other illustration should be marked on the back, distinctly but lightly, with soft pencil with first author's name, figure number, manuscript page number, and the side which is the top.*

If illustrations from published books or periodicals are used, the exact source of each should be included in the figure legend; if these "borrowed" illustrations are copyrighted by others, permission of the copyright holder to reproduce the illustration must be secured by the author.

5. Nomenclature

All pesticides and other subject-matter chemicals should be identified according to *Chemical Abstracts,* with the full chemical name in text in parentheses or brackets the first time a common or trade name is used. If many such names are used, a table of the names and their precise chemical designations should be included as the last table in the manuscript, with a numbered footnote reference to this fact on the first text page of the manuscript.

6. Miscellaneous

Abbreviations. Common units of measurement and other commonly abbreviated terms and designations should be abbreviated as listed below; if any others are used often in a manuscript, they should be written out the first time used, followed by the normal and acceptable abbreviation in parentheses [e.g., Acceptable Daily Intake (ADI), Angstrom (Å), picogram (pg), parts per trillion (ppt)]. Except for inch (in.) and number (no., when followed by a numeral), abbreviations are used without periods. Temperatures should be reported as "°C" or "°F" (e.g., mp 41° to 43°C).

Abbreviations

A	acre	kg	kilogram(s)
bp	boiling point	L	liter(s)
cal	calorie	mp	melting point
cm	centimeter(s)	m	meter(s)
cu	cubic (as in "cu m")	μg	microgram(s)
ft	foot (feet)	μl	microliter(s)
gal	gallon(s)	μm	micrometer(s)
g	gram(s)	mg	milligram(s)
ha	hectare	ml	milliliter(s)
hr	hour(s)	mm	millimeter(s)
in.	inch(es)	m$\underline{M}$	millimolar
id	inside diameter	min	minute(s)

<u>M</u>	molar	lb	pound(s)
mon	month(s)	psi	pounds per square inch
ng	nanogram(s)	rpm	revolutions per minute
nm	nanometer(s) (millimicron)	sec	second(s)
<u>N</u>	normal	sp gr	specific gravity
no.	number(s)	sq	square (as in "sq m")
od	outside diameter	vs.	<u>versus</u>
oz	ounce(s)	wk	week(s)
ppb	parts per billion	wt	weight
ppm	parts per million	yr	year(s)
/	per		

Numbers. All numbers and fractions or decimals are arabic or roman (table numbers only) numerals. Numerals should be used for a series (e.g., "0.5, 1, 5, 10, and 20 days"), for pH values, and for temperatures. When a sentence begins with a number, write it out.

Symbols. Special symbols (e.g., Greek letters) must be identified in the margin, e.g.,

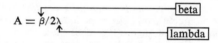

$$A = \beta/2\lambda$$

Percent should be % in text, figures, and tables.

Style and format. The following examples illustrate the style and format to be followed (except for abandonment of periods with abbreviations):

> KAEMMERER, K., and S. BUNTENKÖTTER: The problem of residues in meat of edible domestic animals after application or intake of organophosphate esters. Residue Reviews <u>46</u>, 1 (1973).
> The Chemagro Division Research Staff: Guthion (azinphosmethyl): Organophosphorus insecticide. Residue Reviews <u>51</u>, 123 (1974).

7. Page proof (Galley proof is no longer sent)

Corrected proof must be returned, within two weeks of receipt, to the editor. Author corrections should be *clearly* indicated on proof with soft pencil or with ink and in conformity with the standard "Proofreader's Marks" accompanying each set of proofs. In correcting proof, new or changed words or phrases should be carefully and legibly handprinted (*not* handwritten) in the margins.

8. Reprints

Senior authors receive 30 complimentary reprints of a published article. Additional reprints may be ordered from the publisher at the time the principal author receives the proof.

9. Page charges

There are no page charges, regardless of length of manuscript. However, the cost of alterations (other than corrections of typesetting errors) attributable to authors' changes in the page proof, in excess of ten % of the original composition cost, will be charged to the authors.

If there are questions that are not answered in this leaflet, see any volume of *Residue Reviews* or telephone the Editor (see p. 1 for telephone numbers). Volume 3 (Ebeling) is especially helpful.